ENVIRONMENTAL RESEARCH ADVANCES

URBANIZATION

GLOBAL TRENDS, ROLE OF CLIMATE CHANGE AND EFFECTS ON BIODIVERSITY

ENVIRONMENTAL RESEARCH ADVANCES

Additional books in this series can be found on Nova's website under the Series tab.

Additional e-books in this series can be found on Nova's website under the e-book tab.

ENVIRONMENTAL RESEARCH ADVANCES

URBANIZATION

GLOBAL TRENDS, ROLE OF CLIMATE CHANGE AND EFFECTS ON BIODIVERSITY

HAROLD J. KAUFFMAN
EDITOR

nova publishers
New York

For permission to use material from this book please contact us:
Telephone 631-231-7269; Fax 631-231-8175
Web Site: http://www.novapublishers.com

Library of Congress Cataloging-in-Publication Data

ISBN: 978-1-63117-063-8

Published by Nova Science Publishers, Inc. † New York

CONTENTS

PREFACE

Urbanization is caused by migration. However, an essential difference exists between migration in the past and that in the present. This book focuses on several topics that include a comparison between an informational city and a smart city; urbanization and human impact on biodiversity in Water's Edge Public Spaces (WEPS); a mathematical-modeling approach to urbanization caused by migration; sustaining urbanization in a developing economy; and urbanization and cardiomyopathy vulnerability in an African male cohort.

Chapter 1 – At the end of the twentieth century, with the emergence of communication technologies in urban space, new towns begin to emerge seeking a new way of living through paradigms of flexibility, sustainability, diversity, innovation. These were some of the characteristics of Celebration City, designed and built by Disney Co. and facing a global, differentiated and high income audience, which wished and accepted the values proposed in the design of houses and streets in the urban idealized behavior. Today Celebration City is an urban reality and a model for informational cities. In the second decade of this century, Songdo in South Korea which is under construction entering the Yellow Sea, is a model for Smart and Global Cities. If the American City had proposed to ensure global flow spaces with local spaces equipped with high technology, in the Asian City the relationship between global and local is an endogenous datum somewhat inherent in urban phenomenology, i.e., networks exist because the material space thus demands, and not to take the place of the latter one. The two cities studied show trends that drive the global urban proposals, seeking solutions for the present-future in a world where cities are the main environment and biodiversity gains extended meanings.

Chapter 2 – Public spaces along revitalized urban waterfronts, known as Water's Edge Public Spaces (WEPS), are facing the consequences of rapid urbanization and urban transformation in Asian and Pacific contexts. In the Australian context, these spaces are increasingly being upgraded to meet the requirements of a changing demography and the needs of this slowly growing population. Increasing use of public spaces requires management authorities to investigate considerate design techniques and to make a long term strategic commitment to urban renewal, to enable appropriate future use. To manage the process of urban renewal, transformation and growth, cities need to be designed and retrofitted to be more environmentally friendly, for example, to decarbonise their energy supply and minimise waste in all forms, to protect, sustain and enhance urban biodiversity and to enable ecosystems to flourish. This will provide inhabitants with the basic elements of human well-being in a resource- and energy-efficient manner. Public spaces play a key role in this transformation process, and over time will be crucial to the success of cities.

The impacts of urbanisation are apparent through the study of developed inner-city waterfronts, riverfronts, and ports, especially in Sydney, Hong Kong and Singapore. Locally, the impacts of demographic change and low level population growth are of a different nature and type in the Australian spaces studied. Besides social and economic impacts, the impacts of urbanization are environmental, ecological and biological, and becoming visible in the study of predator-prey models that analyze population numbers over time. To ensure the maintenance and flourishing of flora and fauna, we need a well-considered design process that uses the holistic principles of green urbanism.

This chapter examines biodiversity within three cities(that are being investigated as part of wider research) to determine the forms of design that are most conducive to change by allowing varied species to thrive and maintaining healthy ecological systems within new and existing urban habitats. It is assumed that a species' natural environment is its most favourable habitat, but, in an urbanizing milieu, adaptation that is assisted through design evolution and development that allows flexibility and change, while maintaining stable population states is sought. The variables and parameters to enable biodiversity maintenance and reestablishment should be considered. As a conclusion, this chapter provides key design considerations to preserve biodiversity within Water's Edge Public Spaces in the Asia-Pacific region.

Chapter 3 – Urbanization is caused by migration. However, an essential difference exists between migration in the past and that in the present. Hotelling constructed a mathematical model which can accurately describe the former case. In quantitative sociodynamics, Weidlich and Haag constructed a mathematical model which can accurately describe the latter case. The behaviors of these mathematical models are very different from each other, and the behavioral difference expresses the essential difference between migration in the past and that in the present. By comparing the mathematical models, in the present research the authors prove that the essential difference is caused by the difference between the cost of movement in the past and that in the present.

Chapter 4 – This study models the relationship involving globalization, urbanization, and sustainable development (SD) where SD is defined further as a function of sustainable urbanization. It repositions culture as one of the pillars of SD, underscores SD's operational definition as "complementarity" as against "uncompromising-ness" of efforts to meet the present and the future needs, and advances a performance-based SD index. This study delineates between the two terms "developing economies" and "emerging economies" in favor of the former. Accordingly, it adopts Nigeria as a case study of a developing economy grappling with urbanization (largely due to rural-urban divide). Nigeria is the fastest urbanizing country in Africa based on its population. Conversely, 80% of Nigeria's 50% urban population lives in slums, attributable to Nigeria's housing deficit of over 16 million units, which is increasing annually. Identifying housing as a menace to as well as an indicator of Nigeria's urbanization, this study summarizes the failure of prior efforts on housing in Nigeria into not being CLASS (culturally-acceptable, livable, affordable, sociable, and scalable) as well as not being SMART (specific, measurable, achievable, relevant, and time-bound). Premised on a SMART CLASS solution, it proposes an open-loop urbanization model based on decentralized developments in Nigeria as against the prevailing focus on the capital states. With the increasing volume of foreign direct investment (FDI) in Nigeria and concerns of the benefits thereof, this study also investigates the role of foreign players in sustaining urbanization in Nigeria. Specifically, it underscores China's significant role as a development agent in Nigeria, its leading role in ISO standards on quality management and environmental management, and its exemplary role as the fastest urbanizing country in the world.

Chapter 5 – *Objective:* Vulnerability for hypertension in Africans increases when they are exposed to psychosocial stress or urbanization. Inevitable cultural and lifestyle changes may impact on cardiomyopathy risk markers. Therefore, the authors aimed to assess the relationship between blood pressure, silent ischemia, glycated hemoglobin (HbA1c), QTc dispersion and

ECG Cornell Product left ventricular hypertrophy (LVH) in a cohort of African men.

Methods: The authors included 100 Caucasian and 82 African men, aged 25-60 years, from the Sympathetic activity and Ambulatory Blood Pressure in Africans (SABPA) Study. HbA1c as well as ambulatory blood pressure (BP) and -silent ischemia values were evaluated, compared and correlated. QTc dispersion and Cornell product data were obtained via 12-lead ECG.

Results: African men revealed ($p \leq 0.05$) pre-diabetic status with a mean HbA1c (> 6%). In addition, they revealed increased ambulatory BP, silent ischemia, QTc and LVH values compared to the Caucasian men. In forward stepwise regression analyses [Adjusted R^2 for model, 0.33], ambulatory systolic BP (0.53, 95% CI 0.12 to 0.94) and silent ischemia (0.26, 95% CI 0.06 to 0.46) were associated with LVH in African men.

Conclusion: Systolic BP and silent ischemia were associated with LVH in a cohort of African men. Increased cardiovascular disease risk markers may reveal progressing cardiomyopathy.

In: Urbanization
Editor: Harold J. Kauffman
ISBN: 978-1-63117-063-8

Chapter 1

WALT DISNEY'S CELEBRATION CITY AND SONGDO/SOUTH KOREA: COMPARISON BETWEEN INFORMATIONAL CITY AND SMART CITY, TWO URBAN PARADIGMS FOR THE WORLD

Paulo Celso da Silva*
Professor at University of Sorocaba, São Paulo, Brasil

ABSTRACT

At the end of the twentieth century, with the emergence of communication technologies in urban space, new towns begin to emerge seeking a new way of living through paradigms of flexibility, sustainability, diversity, innovation. These were some of the characteristics of Celebration City, designed and built by Disney Co. and facing a global, differentiated and high income audience, which wished and accepted the values proposed in the design of houses and streets in the urban idealized behavior. Today Celebration City is an urban reality and a model for informational cities. In the second decade of this century, Songdo in South Korea which is under construction entering the Yellow Sea, is a model for Smart and Global Cities. If the American City had proposed to ensure global flow spaces with local spaces equipped with

---

* Corresponding author: Paulo Celso da Silva. E-mail: paulo.silva@prof.uniso.br.

high technology, in the Asian City the relationship between global and local is an endogenous datum somewhat inherent in urban phenomenology, i.e., networks exist because the material space thus demands, and not to take the place of the latter one. The two cities studied show trends that drive the global urban proposals, seeking solutions for the present-future in a world where cities are the main environment and biodiversity gains extended meanings.

INTRODUCTION

Celebration – The Disney City: Technology and Environment

You Don't Have to live here to love it.

When getting to Celebration, the first impression a visitor has is the "atmosphere of an old "unitedstatian city". From the Celebration Realty Inc, an interesting building, with a belvedere from which it is possible to have a full view of the plane extension of the city. This is the experience of a tourist visiting Orlando and the accomplishments of Disney Co.

The *Celebration Avenue* looks endless for someone who walks along it, and it is possible to see one of the houses whose design has some arches similar to the ones in the movie Cinderella. North Village is known for its *Model Homes.* These are houses designed to allure the ones who intend to live enjoying the advantages of technology. The prices are surprising. There are several models for different prices.

The houses, besides their architecture, they are all equipped with Intranet – a local net – and it is possible, for example, to attend the meeting of teachers or parents, at school, in the home computer. It is the future beyond the future, the convenience of an apparent stable reality, in which, living outside Celebration may be risky and unadvisable. Thus, it is impossible not to perceive that the fantasy atmosphere herein is similar to the parks.

Back to Town Center, the visitor can observe – along the wide avenues – that there is no car pollution in this city which insistently tries to present itself as an option for man to live harmonically integrated to landscape. By the way, the landscape here is absolutely cohesive concerning green areas and there is also a golf court, completing the scenery, which used to be swampy. On the way we also found a private club for residents.

At the Town Center bank, the folder given by the estate agent read: "You Don't Have to live here to love it".

The city has been designed to stand for a novelty, it was fully planned, in order to look traditional and at the same time, modern and functional, concerning all possible urban aspects. Its infrastructure – which is differentiated from most American cities – shows it clearly. Demographically speaking, there is another aspect that must be taken into account: The original project was estimated for 20,000 inhabitants occupying 8,000 houses by the end of 10 years (begun in 1995), however, according to 2010 Census[1], the number of inhabitants was 7,427 people, distributed in 3,063 houses.

The first stages of building included: Celebration Village, West Village and Lake Evalyn with 500 home sites, including 76 townhomes, and nearly 700 condominiums. The second stage was North Village, with more than 300 home sites and 300 condominiums, followed by South Village, with over 600 home sites and townhomes and nearly 500 condominiums. Following was East Village and Aquila Reserve with a combined 500 homes and townhomes and 70 condominiums. Artisan Park, Celebration's last residential phase to date, is a 160-acre village built by St. Joe Company and contains more than 600 homes and condominiums, as shown at the website of Celebration[2].

Nevertheless, the uppermost question clearly put is: 'Why did Disney Co. invest in a 4,900-acre area real city, instead of a park? As a matter of fact, it is said that Walt Disney himself had designed, in the 60's, a city for the employees of the parks, based on Victor Gruel's urban proposals; however, his idea ended up being changed to *Epcot Center.*

Thus, the present Disney Co. would be "making its founder's dreams come true"... After all – to carry out this project – *The Celebration Company* counts on strategic partnership with different companies: ATandT, which is responsible for the communication system, computer and telephone equipment for the shops, banks and houses; the Florida Hospital, where there is the *Celebration Health,* which encompasses several medical-hospital services, including the *Fitness Center,* a rehabilitation center, health plans, behavior research center, odontological center, among others; GE Company, with its wide set of electrical devices and light support for houses and hospitals; the *Honeywell Home Control,* which offers systems and services related to comfort, protection and economy for houses (alarm and ventilation systems etc.); *Sun Trust,* a bank service; the *Stetson University,* responsible for the *Celebration School,* a public school for children and teenagers.

---

[1] US. CENSUS 2010. [2013, October, 15]. [Celebration City's residents] Available from: URL: http://www.census.gov/2010census/.

[2] CELEBRATION. [2013, October, 14]. Available from: URL: http://www.celebration.fl.us/town-info/community-profile/.

Besides that, it is also responsible for *The Celebration Teaching Academy,* which develops extension and specialization courses for professors and others who may be interested. The *Celebration Hotel,* with 105 rooms and 10 special rooms – 1920 style, designed by *Owner and Developer* – which was opened.

We have to accept that – even a simple bench in a square, close to an old water fountain – the city impresses us strongly! It can be said in relation to many aspects, because the project is detailed and, thus, we are touched by the magic of the place. In 1998, Mrs. Elaine McHenry, a Brazilian who lives in the city, was interviewed; she said a symptomatic phrase:

> This city was Walt Disney's great dream. Today's Disney only aims money; Walt Disney was a visionary. We live in a paradise, but nothing here relates to a paradise. It is beautiful, well preserved, clean, there is even a police car driving around. Just a little is left from Walt Disney's idea; most of the employees of the park live in *Kissimmee or Osceola County,* in the surroundings of Celebration, because of the lower cost of living, rent and the price of the square meter in the estate market.

But living in the Disney world implies its proper rules. By the end of the 1990 decade, as the city still did not have a clear identity, people used to arrive, expecting to find "their" city. The same resident also said: "As it is something new, each one who came had idealized a city, that is, his/her city. But there is an important detail; it is necessary to adapt to its rules". These rules are based upon four principles that direct the community living[3]:

> Sense of Place
>
> All people have a universal need for stimulating experiences and social interaction. Providing a variety of unique "places" throughout Celebration, where this type of activity would continually occur, was an important Cornerstone behind its creation. "Places" take on many forms at Celebration.
>
> Focus on Technology
>
> From the beginning, Celebration has embraced technology as an essential means of connecting people. Today, every resident has access to a Celebration email address and an electronic identification card. To accommodate future advancements, Celebration is prepared to evolve

[3] CELEBRATION. [2013, October, 14]. Available from: URL: http://www.celebration.fl.us/town-info/community-profile/.

with technology as it changes, helping to ensure that technology will continue to do exactly what it is intended, to bring people together.

Focus on Education

Celebration has placed education at the heart of its vision in the belief that lifelong learning provides an essential key to higher quality life. In 1995 education began at The Learning Academy, a public education system recognized by the Florida Department of Education with a 2000-2001 Five Star School Award. Currently educational opportunities include KinderCare, a Montessori School, Stetson University Center at Celebration, Celebration K-8 and Celebration High School.

All the residents corroborate the quality of living and schooling: "I like living in Celebration, it is like being on vacation, you can walk, go Down Town, to the restaurant. I moved to Celebration because of the school, which is different. Every student is demanded to present a research on the subject. At the end of it, the result is: done or not done". That is what Mrs. Elaine McHenry said, when expressing her opinion.

Besides that, in the beginning of the project, there was a manual named *Frequently Asked Questions, compiled by Kenny Cottrell*[4], in which we find principles of adequate conviviality, besides a compulsory approval to "radical" changes in the houses. These are some examples:

> Every resident who is an owner will be a member of the Celebration Residential Owner Association, and will have to pay for the Association maintenance fare. This money will be used to pay for maintenance services, including *Lakeside Park* and other public areas within its jurisdiction. The Association will keep the landscape and grass in front of the houses, except in *Townhomes*. Private landscape and grass areas will not be kept by the Association.
>
> All the owners of houses located inside the Celebration Community Development District (CCDD) will have to pay yearly fares, in order to cover the cost of building and maintenance of public areas under its control. The fares will be paid through tax invoices in Osceola County. Based on this evaluation, the cost of living in Celebration is the same as in any high level community, provided with similar services. There will be some restrictions, according to the Convention, Restriction and

[4] Frequently Asked Questions, compiled by Kenny Cottrell. Acesso em 03.05.2000. O site http://www.home.ptd.net/~glisman/cele2.htm. Is, now, unavailable.

Condition Declaration. Approving styles and aesthetic matters in Celebration is an attribute to both *Celebration Co.* and the Association.

The official regulation for the use of public spaces was published in October 2013. The headlines of the Celebration News (2013, p.7) read:

> The Celebration Residential Owners - Association (CROA) Board approved the following resolution to the current ID Badge policy: A passholder may bring no more than four (4) non-resident guests outside of their immediate family on any given occasion unless under a scheduled event, private party, etc. The passholder must remain with the guest(s) at all times. In order to avoid conflicts or disappointments please contact Town Hall if you are having a party in one of the parks or playgrounds[5]

Although this city has been well designed in the urban, architectural and technological aspects, so far it has not been fully accepted by the American society. In our research on this matter, some articles show how unsatisfied many people were concerning this enterprise[6].

> Celebration must be considered more a corporate-run subsidiary than a normal city: No municipal power, the association for no estate owner and no free trade; besides a “Regulation Standard Book”, strict and related to maintenance and care of the residents’ “property”. In most of the cases, a city like this is called dictatorship, at Disney it is Celebration. The general hypothesis is that the corporation will make the city, reasonably more efficient, and they think that the representative powers will offer a better living atmosphere. Disney could perceive the economic moment of high income families, who were a little nostalgic and also greatly unsatisfied in their communities. Thus, it is the right moment and the perfect place to trade your products.

It is visible that everything in Celebration has the aspect of a clean hygienic place; its parks are well preserved, visually beautiful and organized, all of them created for Magic Moments. The kind of moment that Walt Disney proposed, which should be enjoyed – at least once – by everyone who is used to visit green leisure areas like this.

[5] CELEBRATION NEWS. [2013, October, 20]. Available from: URL: http://www.celebration.fl.us/wp-content/uploads/2013/10/CELEBRATION-NEWS-OCT-2013-low.pdf.

[6] DISNEY'S UTOPIA. [2013, October, 20]. Available from: URL: http://www.123helpme.com/disneys-utopia-view.asp?id=164640.

These are simple but highly impacting moments for the resident-visitor. For example: If a child's pacifier is dropped on the ground, it is expected an employee of the park to appear and offer a brand new one immediately to the mother, thus, avoiding any possible dissatisfaction of the family.

This environment creates a false impression of "eternal protection" and, somehow it seems that Orlando, Orange, Kissimmee and their nice surroundings are designed to cause this feeling of welfare, safety and comfort.

Michel Serres, in his work, indicates the concept of message and, on page 250, the philosopher states:

> By means of a virtual extension of Geography, fundamental knowledge, because we, living beings, inhabit this world, as fruit trees or animals on a valley, towards to a new cartography, which contains virtual spaces… We do not have, however, any evidence that this prolongation will prolong, one by one: its transitivity is more frequently interrupted than continued.

Here, transitivity is great, each tourist creates his virtual map, the map of his possibilities in this "magic" process of confirming "in loco" this amazing new world. See, see, consume, see, consume… Actually, it is the continuous circle that keeps the permanent feeling that we are alive because we are consuming. Virtual cartography mapped in knick-knack shops, enchanted geography in this Babel Tower of endless consumption.

Again we present one of Serres' statements (1995, p. 72) to complete:

> Here is a novelty in our history: unique and indestructible, universal, the new city keeps us from being hopeful, except in relation to itself and its performances; nobody enters unless they have full access everywhere. Well, previously, The Paradise was perceived and imagined as the place of inaccessible Hope. Neither the most sacred tradition, nor Dante's geniality, could predict that one day; man would be permeating all the places with technology similar to gold twigs.

This is an apparent quality of Celebration, however, it is foremost a social process being highlighted. Previously, it was shown and emerged the matter of paradise, according to Mrs. Elaine's statements, a resident of the model-place. Thus, the discussion on social processes is retaken, having the quality of place as measuring criterion. It is about living, feeling, moving, to have access in Celebration, that is what really matters.

Celebration is what it is, an international capital message. The current living at the houses completes the networks working in cybernetic and mechanical fluxes – coaxial cables, streets, avenues, optical fiber. It is the apparent human absence or an inhuman city.

Its economical-financial complex center encompasses global capital companies. AT and T, GE Co., Sun Trust, Honey Home Control, Florida Hospital, all of them located at Business Park. These are apparently harmonic conviviality of the capital striving for the capital.

The residents united at their 'Association" complete this new "contract" of tomorrow's citizenship or a new social order, clearly based on networks, as Manuel Castells [*1999: 505*]. Keeping pace with the Spanish sociologist's thought, it is the architecture of nudeness and of the silence message that intends to say nothing (1999: 445).

This order looks strange to people outside, it is a kind of order that makes one feels liberty-castrated, for the simple imposition of the network existence itself.

In itself, the internal debate about Celebration ranks the city among the ones in which information is the main product, labor value and income producer. But only this would not be sufficient to make the city informational; it would be at the most a polemic city with advanced technology. However, Celebration is admitted as another possibility of gaining in the flexible accumulation world.

Conceived and in process of making, within new capitalist parameters, Celebration unites the virtual city with its Intranet and its intranet living, to the dynamics of the informational city, a situation that is generated by the technological capital, accumulated chiefly on the service field.

The conception of the city itself is related to flexibility, since the first question that emerges is why Disney Co. has not created another park instead of a city?

Then, according to the fordist logic, the obvious would be investing in entertainment, but Disney Co., within a "consortium", prefers to call it "Strategic Alliance"[7], with several companies that put into practice a "dream" of the visionary Walt Disney himself, believing, first, in the force of the brand

[7] The Celebration Company has formed strategic alliances with a variety of prominent companies to help bring the latest products, services and technologies to Celebration. Relationships evolved during the last four years of planning. More alliances are yet to be forged. Alliance relationship encompass businesses in the building products, communications, health care, education, financial and civic areas. The following companies are discussing their participations in Celebration with Celebration Company. Fax received on 08.31.98, The Celebration Company, 200 Celebration Place, Celebration, Florida, 34747 F.(407)9391999.

in which the Disney concept of entertainment has become; besides that, on the impact of the technology on people's lives; then on 10/18th/1995, US Today published, on page 5B – just in the beginning of building – a report asking: "Will it work for the late 20th century? A picture-perfect town, but will it work?

Disney Co. knew how to play with people's imaginations, being able even to transmit the marketing message of an American company that is specially regarded to the American values, according to which, having a good standard of housing is beyond a mere concept of something conventional, and managed to implement an idea of future associated to tradition. Therefore, values that have a price and they are not so high, when the credibility is guaranteed and when it comes to talk about safety, comfort and stability.

Thus, the creation of Celebration is an attempt, so far apparently successful, to create also an identity for its inhabitants, once they are from many different places, and have, sometimes, different conceptions of city.

Concerning the matter of tradition allied to technology, Celebration, in its design from XIX and XX century, depicts the postmodernist possibility of simulacrum. This, on practice, presents traces of a more perfect architecture, implanted on XIX century and reproduced on XIX – XX centuries, proposing such characteristics as a solution for the urbanization of XXI century.

It is the image of non-similarity itself. From the medieval concept of burg to a city that preserves traditions without giving up the technology, as an instrument of comfort and integration between man and this new virtual reality.

It is also heterotopy: each house, each individual as a possible world, fragmented, living in an "impossible" space. And it is impossible within the fordist strictness of an industrial world in quick and constant transformation. Therefore, Celebration slowly incorporates the speed of the information processing, and then, they manage to be awarded by a voracious estate market which adapts itself to flexibility, keeping clear the thesis of permanent renewing: The National Association of Homebuilders, in 1997, declared Celebration as The Best New Community in America, and the city was given the Best in American Living Award. It was a recognition, because it took them three years, though previously, it had been predicted ten years.

But machines are not the only things that change, it is the form of connection to and through machines that make new and possible countless technological interactive processes to any citizen of this city, programmed to work. Another way for both to understand and see the world – here, mixed between private and public – that the "public power" offers to citizenry.

This is another important characteristic of Celebration that is usually addressed many questioning: who rules Celebration, Disney or the chosen rulers?

The city is not independent, it is part of Osceola County and it was administered by people who made part of a consortium of companies in a partnership with The Association of Owners which, expired the ten-year term of the project, transfer it fully to Osceola County. Thus, it was created other way of directing a city, uniting private initiative, popular participation and public power, as a manner to seek solutions for the problems and contradictions that are so evident in a conventional city.

Thereby, trying to understand Celebration as an ordinary standard city is incurring into an analytical error. It is a new proposal, and it is supposed to be understood like that. It is clear that such a conception of city is far from being a terrestrial "paradise"; the city, which is still being made, suggests new styles of urban design and planning.

Labor and socio-spatial relations that are based on the fordist flexibility gained a new dimension due to the association of three social agents that could have antagonistic interests. What was finally shown were some different – but not antagonistic – interests, certain aspects susceptible to be directed aiming the continuity of capital accumulation, as well as a state of governance and satisfaction of the consumer-citizen.

Then, there is a new administrative reality: different forms of gaining joined to different forms of public administration. If fordism could not be hegemonic, in its beginning, here, flexibility is – still – punctual, it is an experienced to be taken inwardly, maybe in a shorter term than the previous model. There are many different things in risk and, after all, theoretically, the society is not determined by technology, actually, the society focus technology from its proper economic condition, facing the availability of the current financial reality.

The technological clip made here, now, is merely theoretical; as it is not possible technology without society, it is obviously not possible society without its technology. Milton Santos has been emphasizing this social dialectics for a long time, a period that was agreed to name technical-scientific-informational. From this perspective, Celebration is a movement of the spatial structuring within this period. Concretely speaking, from the mid of the 90's, when it was really possible to start building the city, everything was under control; all the material demands were being satisfied.

Today, the informational technology applied in Celebration create new projects of living; one example is the Celebration School, which counts on the support of the available technological apparatus that is offered to all its students as a new didactical-pedagogical proposal, in partnership with the Stetson University Center, located on 851 Celebration Avenue, which offered, all year long at the Teaching and Learning Institutes, some courses for teaching improvement, like Innovative Uses of New Technologies: Classroom Applications, Teacher Rejuvenation and, for 2013 there are other proposals like the Stetson Lifelong, which is expected to offer environmental courses like 'Alligators, Snakes, and More – Oh My! or Gardens of the World[8]. However, according to Ross (1999, p. 73) who:

> ...also took part as a volunteer teacher in the educational project, in which he was privileged for having perceived other great conflict source. Little by little, many parents began to question the new method introduced to their children; many points were criticized: it was given some emphasis on classical matters like Math and Science; it is desirable a more disciplined behavior; it is suggested to create a competitive atmosphere to contribute to performance; and as usual, about grades, what could probably, in the future, to impede anyone to enter the university.
>
> At first, the school administration reacted, telling the ones who were unsatisfied, that they could look for other schools, what was somewhat strange, once it was a public school; but besides that, it was also because they had spent a lot of money to buy the houses, and now, they could no longer afford putting their children to study in private schools. Almost one year later, the group that had been claiming for a more traditional teaching style gained new admirers and the school was forced to review its position, concerning many aspects, let al.one the construction of another public school in the area.
>
> Little by little, the reality is that, instead of the emergency of a community as it was previously idealized by its administrators, there is a united group, foremost as an answer, as it was seen, to the real adversities related to their most immediate interests. Such ties, and others that are more informal, overlapped the efforts of the administrators to promote a community feeling, through "block parties" organized by professionals directed by the Disney, in order to create a kind of homogeneous identity, without any "internal fractures".

8 Stetson Lifelong. [2013, October, 15] Available from: URL: http://www.stetson.edu/administration/lifelong-learning/media/lifelong-learning-catalog-celebration.pdf.

Another segment whose style of living was altered by technology and flexibility is health, with the Celebration Health, a technological research and medical service complex joined to the Florida Hospital[9], offering health services which include medical plans, aesthetic institute, women center, behavior center, rehabilitation center and others. It is not simply a hospital, it is a research center with academic production and practice addressed to the pharmaceutical market. The propaganda of Celebration Health says: "The user's really friendly technology" and "the future of health care" or even "spirit, mind and body treatment"

The Celebration Hotel is another enterprise predicted to open in the fall of 1999. It is a four-star hotel, with 115 rooms in three stories, located on 700 Bloom street; it has the simplicity of 1920's Florida design and also a space for conventions and meetings, with a room for about 100-125 people.

Thus, Celebration, this retro-futuristic city, corroborates the main activities of the region where it was established: Tourism/Services, Building and Hospital Services[10].

Nevertheless, it is suggestive to ask: How will Celebration age?

The message is being constantly updated through flux-supports, all of them surpassing-surpassed, I insist, in the space, not in the time.

Nevertheless, our issue is not temporal, it is spatial. Therefore, aging in the space, when there is the possibility to upgrade fluxes, is a datum, not exaggeratedly speaking, out of question. The message indicates new networks that are constantly formed from new possible relations *Ad instar.*

This is the similarity on the difference, it is a dialectics that is spread all over the world, making all the places just another one, however, unique. In the intense postmodernism routine, the show happens on the streets and on the networks (HARVEY, 1989:57). The streets, now, are no longer mere traffic networks for cars and vehicles, but also for message, information. These are networks, now streets, through which information, medicine, medical and financial operations that heal, prescribe the next step of the markets.

These are varied experiences that point to the informational society and its new parameters and challenges. Routinely, the challenge is to overcome the obstacles imposed by the invisibility of these Networks-streets-networks.

---

[9] Celebration Health. [2013, October, 09]. Available from: URL: http://www.celebration health.com.

[10] According to information OSCEOLA COUNTY PLANNING DEPARTAMENT. [2013, October, 15] Available from: URL: http://www.osceola.org/. E-mail Address: oscear@ magicnet.net.

SONGDO – International Business District

Since I live here I can walk to hang out with mine.
Songdo brought me closer to my neighbours

Mrs. Kwon, a resident of Songdo

While the project of Songdo was estimated, in 1990 decade, in 2,5 billion dollars, along the period of ten years that separates its designing until the its final, Songdo Smart city will cost, for an equivalent period of ten years, around 35 and 40 billion dollars, being among the most expensive projects of its kind. It is suggestive to highlight, that the form of the city will not be given only by the material – which, by the way, will be the most modern and expensive one – but all the concept of this city differs from the *unitedstatian* city.

The first city can be considered informational, on the sense proposed by Castells (1995 p. 425), as the one that joins high technology and tradition, and is addressed to the search for community living, articulating local and global fluxes. Songdo is conceived as Smart City, a city where the living experience is intended to be complete, using high technology as a routine ease tool. Inhabit, living, working, having and seeing the world are in the same place, that is, local and global fluxes are mixing elements of the same process. Taking an already known image from the architecture, the Smartcities resemble the idea of lofts, where living, working etc. can be done occupying and changing spaces. However, the similarity or image is restricted to that.

This South Korean Smart city is located in Incheon, an important city and South Korean transportation center, with some ports and the main airport, besides its outstanding role in the Free Economy Zone in Busan-Jinhae, which is 40 miles from Seoul. It is among the one hundred of intelligent cities that are being built around the world, as indicates by Stan Gale, Chairman to Gale International.

The city is set in a 1500 acre-area, taking part of Incheon seashore, entering the sea, and its first part was approved in 2003 and finished in 2009. The project includes commercial and office space, residencies, shops, hotels, as well as civic and cultural installations. In the proposal, 75,000 people are expected to occupy an area of 6,000 km^2 and, by 2016, more than 400 new buildings will exist, including the tallest building in South Korea. The city will be fully equipped with optical fiber, which will permit to control everything from home or to have a video interaction from anywhere in the city.

Nevertheless, it must be highlighted that in Seoul, is is already possible to access high speed wi-fi, in the subway, allowing to send an e-mail or to watch videos peacefully. Other technological advantages still include electronic panels indicating the time of trains and buses. Thereby, the south Koreans are used to making use of technology, but Songdo intends to indicate a new form to live the city, now with sustainability and economy of resources, besides recycling a lot of scraps.

This can be verified, observing the networks of scrap disposing that circulate, going from house to house. Then the material is taken to a central station, which it is processed automatically in a hygienic process. In the city, 10% of its total area is destined to preservation of Nature, so they made a big park, inspired on the Central Park in New York. It is a 100-acre area with a sea channel (seawater canal); and 25 km of cycle road to offer an option of faster and clean moving.

The idea is the connectivity between places and, chiefly people; it is the differing factor that is most cited by designers and builders of the city, as confirmed by Summer Scott, when he stated that, maybe, the greatest instrument for urban living, in Songdo, is the technology that connects everybody, independently on where they are, through a video; Stan Gale, vice-president to Gale International completes the great concept to Songdo: "We are a compact city, we are a green city".

To reach the objectives proposed in this designing and building, the city was divided into several sectors of houses, it is necessary to present them:

The First # World (first building delivered in the city in 2009, with 1596 residential units);

Central Park I (729 Residential Units, Ground Level Retail, Completed: November 2010);

Central Park (II 632 Units, Retail on Levels 1-3, Completed: August 2011);

Central Park III; (380 Residential Units, Estimated Completion: The first half of 2016);

Canal Walk (345 Officetel Units, 993 Parking Spaces, All Units Sold Out (2013), Completion: October 2009);

XI Harborview (1,069 Residentail Units, 45 Retail Units, Two 33 Storey Towers and Two 41 Storey Towers with Two Underground Levels, 2,221 Car Parking Lot, All Units Sold Out in 5 Days, Completed: February 2011);

Songdo Prugio (593 Residential Units, 17 Retail Units, Completed: March 2011);

Songdo Prugio Harborview II (Block D18, 662 Residential Units, Ground Level Retail 1,329 Car Parking Garage, Estimated Completion: November 2013);

The # Harborview (845 Residential Units, 12 Buildings Ranging from 13 to 33 Storeys

90% of Units Sold Out Immediately, Completed: April 2011);

The # Harborview II (548 Residential Units, 10 Retail Units, Eight Buildings Ranging from 14 to 38 Storeys with Two Levels Underground, 1,124 Car Parking Garage, Completion: May 2012);

The # Expo (1,337 Residential Units, Completed: September 2009);

The # Green Avenue (Block D7-1/8, 1,014 Residential Units, 21 Buildings Ranging from 12 to 28 Storeys, 2,044 Car Parking Garage, Estimated Completion: June 2012);

Block D11/16 (1,196 Residential Units, Completion: April 2013);

Block D17 (298 Residential Units, 601 Car Parking Garage, Estimated Completion: November 2013):

Block F 21/22/23-1 (1,654 Residential Units, Estimated Completion: July 2014)[11]

There is still an important relation housing-labor, it is important in Songdo, to reinforce the idea of intelligent city, this is clearly shown in the rational division of the productive sectors:

NEATT

Block A2

68 Floor Tower with Office, Hotel, and Residences

Floors: B3F-68F

Building Height: 1,000 Ft

Office: Levels 2, 6-24

Hotel: 204 Rooms on Levels 25-36

Residences: 200 Units on Levels 37-62

Observatory and Restaurant: Levels 63-65

Direct access to Subway Station

[11] Songdolifestyle. [2013, October, 15] Available from: URL http://www.songdo.com/ songdo-international-business-district/the-city/lifestyle.aspx.

Direct connection to the Riverstone Shopping Center and Central Park via Skybridge
Security Guard Desk at the Public Entrance Lobby
18 passenger lifts serving the office towers over 3 zones
Estimated Completion: 2013
Buildable Area: 2.2M SF/202,420 SM
Architect: KPF/Heerim
Contractor: Daewoo/POSCO E and C
Pursuing LEED CS Certification. Targeting Silver Rating.
Available for Lease

International Plaza
Block F3-1/3
5 Buildings with Office, Officetel, Retail, and Public Parking
Will House uLife Building and Sustainable international Business Center
Buildable Area: 2.5 M SF/223,262 SM (Office, Officetel, Garage)
Estimated Phase I (F3-1/3) Completion: Apr. 2014
Architect: KPF/Baum/Dongil
Contractor: POSCO E and C
Buildings Aiming for LEED Certification. Targeting CS Silver and CS Gold Ratings.
Available for Joint Venture or Lease

Gateway Center
Block F6/7
5 Buildings with Commercial, Officetel and Retail Facilities
3,480 Car Parking Garage
Buildable Area: 5 M SF/465,000 SM
Architect: KlingStubbins/Heerim
Contractor: TBD
Pursuing LEED Certification. Targeting CS Silver and NC Certified Ratings.
Available for Joint Venture or Lease

Songdo Convensia
Block A2
Opened: October 7, 2008
34 acre site

3 Exhibition Halls with 5,000 people Capacity, 23 Conference Rooms, 3 Ballrooms
600 Car Parking Garage
Buildable Area: 583,000 SF/54,000 SM
Largest Column-Free Space in Asia
Donated to the City of Incheon
Operated by Incheon Tourism Organization
Architect: KPF/BAUM
Contractor: POSCO E and C
Pursuing LEED for NC Certification

POSCO E and C Headquarters
Block F1
Two 39 Storey Towers with 5 Underground Levels
1,124 Car Parking Garage (with 35 Handicapped Parking Spaces)
Buildable Area: 1.6 M SF/150,000 SM
3,200 people will work when completed
Completed: June 2010
Occupier: POSCO E and C
Architect: Nikken/Sekkei/POSAC/Gansam
Contractor: POSCO E and C

Centroad
Block B4
1.5M SF/142,077 SM Office Space, 264 Officetel Units, 294 Retail Units
1,678 Car Parking Garage
Buildable Area: 2.1 M SF/201,952 SM
Estimated Completion: Oct 2011
Architect: Heerim/Gansam
Contractor: POSCO E and C
Developer: SD Advisor
Pursuing LEED for NC Certification.

Daeduk Office
Block B5
1,635,587 SF/ 151,951 SM Office Space, 156,206 SF/14,512 SM Retail Space, 505,441 SF/ 46,957SM Officetel Space with 194 Units
1,694 Car Parking Garage
Buildable Area: 2.3 M SF/213,421 SM

Estimated Completion: Mar. 2014
Architect: Sangji
Contractor: Daeduk E and C
Developer: Songdo Best
Pursuing LEED for NC Certification

International Business Square
Block C8-2
Mixed-Use Building With 1,197,894 SF/111,288 SM Office Space and 119,953 SF/11,144SM Retail Space
Buildable Area: 1.3M SF/123,200 SM
Completed: July 2011
Architect: Heerim
Contractor: Daewoo E and C
Pursuing LEED for NC Certification.

A1 Office
Commercial Building
Buildable Area: 1.2M SF/111,461 SM
Contractor: POSCO E and C
Pursuing LEED CS Certification. Targeting CS Silver Rating.

Commercial Development Opportunities

Lease, Joint Venture, Build-to-Suit distributed in several blocks[12]

Presented like that, the city of Songdo is, at the same time, a reality, because its first phase is finished and there are residents already inhabiting the city and they present it to the world through the virtual resource of social networks.

It is seen, for example, a smiling and happy resident, Evren Rey, at Youtube, in 2011, who shows to the spectators, some emblematic points of the city, in 2:02 min., giving a special emphasis at the Central Park and the International School.

Jeff Mercer, Headmaster to Chadwick International School, describes it like that: "The design and the purpose of Chadwick International School are to educate the children of the businessmen in new Songdo city. The k-12 private school is linked to its founding campus in Palos Verdes, California. Both

[12] Working Songdo. [2013, October, 16] Available from: URL: http://www.songdo.com/ songdo-international-business-district/the-city/working.aspx.

schools use technology in every classroom, to improve the learning experience".

Conclusion

Here it was presented, in general lines, two urban experiences conceptually different, both in the form of creating and organizing space or in the way of living. Celebration is an informational city, whose purpose is to offer to the consumer-resident, an aesthetic alternative in the union between technology and tradition. Big groups of companies offer and sell the most up-to-date products and modern apparatus that are expected both for home use and to ease living.

The Asian city, for its turn, offers the concept of Smart City, where technology is not offered in itself, but, connected to the resident's intelligence. It completes a relation that, in its limit, will be adding quite a lot to the lived territory. The economy of energy, water, time, money results in an interesting model of space intervention, chiefly when we verify that it was designed upon a sea earthwork. Townsend (2013 p. 15) defines an intelligent city, as urban ambiance, where information technology is combined to the infrastructure, architecture, routine objects, and even our bodies, to deal with social, economic and environmental problems.

Thus, even if the concepts are different, both cities have technology, as a differing factor to be or dream of being a paradigm to urban planning. In both cases, the general administrators and designers of the project – when they discuss about community living, the importance of having physical spaces for conviviality, meeting people and enjoy leisure – usually use relativistic words when they mention technology.

It is also currently admitted in the discourses, the assurance that, despite all the technology created and applied to the cities, all models are unique and satisfy a specific necessity, in terms of reality and built spaces, producing distinguished results. Thus, there is no standard ready model, which is adequate to apply in other cities across the world. By the way, the future, while an open project, will result unique to each one of them, according to what is set as a challenge to be overcome or in the essence of the city itself, recognizing and valorizing some aspects of its history, as Celebration did, by rebuilding an architectural identity easily recognizable to the average unitedstatian. Songdo has a different position, it is clear that the intention is to be the pioneer in the use of advanced technology as an easing tool to urban living; and at the same time, environment preservation/conservation is the element that provides it.

It seems correct to state that, in both cases, the complexity exists and surpasses greatly, the space selling market discourses that are transmitted.

Similarly to other cities, these ones also have their contradictions that must be cleared in the immediate routine.

Making a poor comparison, in the beginning of the XX century, several places in the world were benefitted, when energy and telephone systems were implanted. This is an informational/communicational network that changed the understanding of the daily routine. Finally, it is possible to produce at any time during the day, it would be possible for the companies, to have a 24-hour working period and the workers managed to adapt – but not without consequences – to this new model of living. The city was the important stage for these changes, or maybe revolutions would be a more precise word to define it.

Since the beginning of the XXI century, a great variety of initiatives have been implanted across the world, backed by informational technologies applied to urban space. The cities become, again, revolution stages on the style of living in the world. Both Celebration and Songdo indicate manners for these revolutions operate, supported by big companies whose investments are heavy. In Songdo, electric cars have specific recharge terminals, at the same time when MIT (EUA) is developing Hiriko (a "folding" electric car), which has already been sold in Europe. Certainly, this will not be the only car in this concept, but the city is already prepared to it and to the others that must be coming.

Perhaps in Celebration, so far, there are no bottles with wi-fi sensors, providing a discount to the user, every time he disposes it in the right garbage container. But one lesson technology has taught us – whose application is adequate to the city designers – the fact that one technology is not supposed to eliminate another one. There is the radio and the TV, and also radio and TV joined; one is not expected to nullify the other. Concerning the cities it is the same; there will be Celebration and its similar ones, as well as, Songdo and a set of other Smart Cities.

References

Castells, M. (1999) – A sociedade em Rede – A era da informação: Economia, Sociedade e Cultura. São Paulo: Paz e Terra, 1999 English version

CASTELLS, Manuel. *The Information Age: Economy, Society and Culture Vol. I.* Cambridge, MA; Oxford, UK: Blackwell.

Celebration. [2013, October, 14]. Available from: URL: http://www.celebration.fl.us/town-info/community-profile/.

Celebration Company. The Official Website. [2005, May, 20] Available from: URL: http://www.celebrationfl.com.

Celebration Health. [2013, October, 09]. Available from: URL: http://www.celebrationhealth.com.

Celebration News. [2013, October, 20]. Available from: URL: http://www.celebration.fl.us/wp-content/uploads/2013/10/CELEBRATION-NEWS-OCT-2013-low.pdf.

Disney's Utopia. [2013, October, 20]. Available from: URL: http://www.123helpme.com/disneys-utopia-view.asp?id=164640.

Frantz, D. and Collins, C. (2000) *Celebration, US: living in Disney's brave new town.* New York, Henry Holt and Company.

Frequently Asked Questions, compiled by Kenny Cottrell [2000, March, 05]. Available from: URL: http://www.home.ptd.net/~glisman/ cele2.htm. Is, now, unavailable.

Gruen, V. (1965). *The heart of our cities: The urban crisis: diagnosis and cure*, Thames and Hudson, London.

Harvey, D. (1989) *Condição Pós moderna*. São Paulo: Loyola. English version Harvey, David. Post modern conditions., Wiley-Blackwell, 1992.

Mitchell, W. J. and Casalegn, F. (2008). *Connected Sustainable Cities.* MIT Mobile Experience Lab Publishing.

Osceola County Planning Departament. [2013, October, 15] Available from: URL: http://www.osceola.org/. E-mail Address: oscear@ magicnet.net.

Ross, A. (1999). *The Celebration Chronicles: Life, Liberty, and The Pursuit of Property Value in Disney's New Town*, New York: Ballantine.

Serres, M. (1995). A lenda dos Anjos. São Paulo: Aleph, 1995. English version SERRES, M (1995). *Angels, The modern Myth.* Flammarion; First UK edition.

Songdoibd. [2013, October, 15] Available from: URL: http://www.songdo.com.

Songdolifestyle. [2013, October, 15] Available from: URL: http://www.songdo.com/songdo-international-business-district/the-city/lifestyle.aspx.

Songdo Working. [2013, October, 16] Available from: URL: http://www.songdo.com/songdo-international-business-district/the-city/working.aspx.

Stetson Education. [2013, October, 15] Available from: URL: http://www.stetson.edu/celebration.

Stetson Lifelong. [2013, October, 15] Available from: URL: http://www.stetson.edu/administration/lifelong-learning/media/lifelong-learning-catalog-celebration.pdf.

Townsend, A. M. (2013). *Smart Cities: Big Data, Civic Hackers, and the Quest for a New Utopia.* W. W. Norton and Company; 1 edition.

US CENSUS 2010. [2013, October, 15]. [Celebration City's residents] Available from: URL: http://www.census.gov/2010census/.

Williamson, L. Tomorrow's cities: Just how smart is Songdo? Available from: URL: http://www.bbc.co.uk/news/technology-23757738 Access in 15.10. 2013.

Wilson, C. (1995). Mickey Builds a Town: Celebration Puts Disney in Reality's Realm. *US Today*. 18. October 1995: 01A, 5B. [2013, October, 15] Also available from: URL http://www.sjsu.edu/faculty/wooda/usacelebration.html.

In: Urbanization
Editor: Harold J. Kauffman
ISBN: 978-1-63117-063-8

Chapter 2

URBANIZATION AND HUMAN IMPACT ON BIODIVERSITY IN WATER'S EDGE PUBLIC SPACES: CASE STUDIES FROM SYDNEY, HONG KONG AND SINGAPORE

Mabel John[1,2] and Steffen Lehmann[1,2]
[1]Zero Waste Research Centre for Sustainable Design and Behaviour,
[2]China-Australia Centre for Sustainable Urban Development,
School of Art, Architecture and Design,
University of South Australia, Australia

ABSTRACT

Public spaces along revitalized urban waterfronts, known as Water's Edge Public Spaces (WEPS), are facing the consequences of rapid urbanization and urban transformation in Asian and Pacific contexts. In the Australian context, these spaces are increasingly being upgraded to meet the requirements of a changing demography and the needs of this slowly growing population. Increasing use of public spaces requires management authorities to investigate considerate design techniques and to make a long term strategic commitment to urban renewal, to enable appropriate future use. To manage the process of urban renewal, transformation and growth, cities need to be designed and retrofitted to be more environmentally friendly, for example, to decarbonise their energy supply and minimise waste in all forms, to protect, sustain and enhance urban biodiversity and to enable ecosystems to flourish. This will provide inhabitants with the basic elements of human well-being in a resource- and energy-efficient manner. Public spaces play a key role in this transformation process, and over time will be crucial to the success of cities.

The impacts of urbanisation are apparent through the study of developed inner-city waterfronts, riverfronts, and ports, especially in Sydney, Hong Kong and Singapore. Locally, the impacts of demographic change and low level population growth are of a different nature and type in the Australian spaces studied. Besides social and economic impacts, the impacts of urbanization are environmental, ecological and biological, and becoming visible in the study of predator-prey models that analyse population numbers over time. To ensure the maintenance and flourishing of flora and fauna, we need a well-considered design process that uses the holistic principles of green urbanism.

This chapter examines biodiversity within three cities(that are being investigated as part of wider research) to determine the forms of design that are most conducive to change by allowing varied species to thrive and maintaining healthy ecological systems within new and existing urban habitats. It is assumed that a species' natural environment is its most favourable habitat, but, in an urbanizing milieu, adaptation that is assisted through design evolution and development that allows flexibility and change, while maintaining stable population states is sought. The variables and parameters to enable biodiversity maintenance and reestablishment should be considered. As a conclusion, this chapter provides key design considerations to preserve biodiversity within Water's Edge Public Spaces in the Asia-Pacific region.

Keywords: Urban waterfront, Green Urbanism, bio-mimicry, rapid urbanisation, biodiversity, cities in the Asia-Pacific region, Water's Edge Public Spaces (WEPS)

1. INTRODUCTION

Cities are constantly transforming. All cities are inherently evolutionary in nature, always fluctuating and morphing, whether the change is positive or negative. It is difficult to examine urban change without considering biodiversity, life expectancy, fitness, environmental quality, social wellbeing and the economic state. A number of indices have tried to quantify such changes and provide measures of urban livability. However, with the impact of globalization, climate change and the urgency of global warming, achieving sustainable urban development has become significantly more urgent and complex. While governments choose directions and pathways, it is broadly accepted that environmental degradation jeopardizes quality of life, for instance, sea levels may change abruptly, leading to a range of associated phenomena. At the same time, there is more and more competition between cities to attract investment and a skilled workforce. In better functioning cities, public spaces need to fulfill a series of expectations and be more than just attractive meeting places; for instance, they should be a key focus for the city, draw tourists, allow socializing, generate energy, collect and store storm water, demonstrate the use of latest technologies, recycle waste, include meaningful public art, and so on. In all this, urban design still warrants a very high priority as it sets out the framework for future development at an early stage and remains central to successful low carbon outcomes (UN ESCAP/Lehmann & Thornton, 2013), in which public space, density and infrastructure have a strong interrelationship. Importantly, designs of such spaces should embrace and enhance existing environmental connections and biodiversity.

The ambitious ultimate goal of these renewed public spaces is self-sufficiency, systemic autonomy and energy generation. Rather than being places merely to meet, eat and socialize, Water's Edge Public Spaces should perform integrated roles and be a part of various urban sectors and systems. They need to operate as part of:

- the energy sector, generating more power than they need to operate;
- the waste sector, collecting and recycling waste;
- the water sector, collecting and purifying water;
- the social system, delivering community benefits, with socially restorative qualities. by facilitating intergenerational and overall exchange;
- the technological system, embracing technological advancements to create learning and connecting points;
- the economic system, generating jobs and providing avenues for expenditure and growth;
- the political system, attaining prominence and portraying esteemed local and global community values; and,
- the environmental system, delivering local and globally linked high-performance self-regulating green spaces.

The integration of trees, shrubs, and flora in Water's Edge Public Space is particularly important in helping to keep the urban built environment cool, because buildings and pavements increase heat absorption and reflection (the urban heat island effect). This integration of various elements to improve public spaces should not be disjoint and disconnected from the original flora and fauna of the space; it needs to enhance the existing environmental qualities by supporting, adding life span and vibrancy to the existing species.

Water level changes (in waterfronts, riverfronts and ports) are an important consideration in terms of public space functioning, especially in terms of disaster mitigation. These changes are based on a number of factors, some of them natural, some inter-related and others man-made. Analysing the causes of water level fluctuations has led to various assertions, many justifiable and some debated with vigour. While this chapter does not focus on water level changes, the authors consider that design should be flexible enough to mitigate disasters caused by water level fluctuations (encouraging floating platforms and clever flexible material use and engineering techniques). Meanwhile, it is understood that access to larger economies provides better opportunities for work, and more people continue to move to globalcities, with hopes and dreams of a better future. A large population is now becoming urban, mobile and connected through technology. In analysing technology use in WEPS, especially considering the Singapore Mass Rapid Transit (MRT), it was found that at least seven out of ten users used mobile phones or headphones while travelling. The urban population is becoming increasingly reliant on mobiles and electronic devices. However, this population is also focused on selfish pursuits, not understanding the impact of their choices and behaviour. Much of this is a consequence of consumerism and image consciousness, but some of it is a result of under-education, 'green-faking' and short-term thinking. The majority of users of public spaces in the global settings of Sydney, Hong Kong and Singapore disconnect from their immediate environments in using mobile technologies, and focus on a specific list of contacts that are available to them. Lowering the possibility of random and chance encounters in physical public spaces and improving these encounters on a virtual sphere, therefore, the nature of interactions is changing. The mobile links that allow for interaction are an associated part of this change, but currently only active seekers of mobile interaction use such technologies to meet like-minded individuals, and to make larger networks based on virtual connectedness.

Emerging Water's Edge Public Spaces have distinct elements. Designing with consideration of these elements allows users the choice of maintaining environmental connections in new spaces if an interdisciplinary design team is used. This is one of those (whether real, perceived or fake) resource intensive tasks where the system cannot yet actually be self-sufficient. WEPS can and should be part of the urban ecosystem to push these public spaces towards self-sufficiency. This chapter focuses on biodiversity in nine selected WEPS in the Asia-Pacific region, and considers whether the re-design or renewal of these spaces should ensure the growth/maintenance of biodiversity and stable species (population) states over time. The spaces analysed are Circular Quay, Walsh Bay and Barangaroo in Sydney; Boat Quay, Marina Bay Promenade and Gardens by the Bay in Singapore; and Star Ferry Pier, Avenue of Stars and Victoria Harbor in Hong Kong. Self-sufficiency and autonomy over generations (for flora and fauna) is supported largely through biodiversity, and architectural principles must reach towards this goal.

2. Changes in Post-Industrial Working Harbours Since the 1980s

The current rate of human impact on the environment, if continued unabated, will have a continued and profoundly irreversible effect on biodiversity (Sala et al., 2000). In parallel, the relationships between working harbours and their city are (like cities themselves) complex and ever-changing. Prior to the establishment of working harbours, biodiversity existed in these spaces. Human presence led to development, which came at the cost of habitat removal or, often, species movement, dispersal or extinction. Since the 1980s, i.e. after port establishment and intensive use, while these areas were originally more connected and openly accessible by members of the public, waterfronts are now becoming renewed mixed use recreational areas, distanced from working ports that contain bulk shipping storage container spaces. The global economic downturn has seen a slowdown in working port activities, reducing use of a commercial nature, and working ports are now being used for different purposes. This can be seen as a further distancing of the city and its recreational activities from working ports in the past five to ten years. This separation of the working port from recreational/mixed use areas of leisure and landmark tourist attraction causes demarcation of uses, but also emphasises individual value in a space i.e. while a number of individuals would previously barter in a certain specific area, the individuals may now be scattered in different cities, but may use the same phone line, mobile application or online/ media to connect, allowing flexibility for them spatially so long as they ensure trading and business relationships continue.

Post-industrial working harbours are now cleaner than a decade ago (when they were working ports) in the Asia-Pacific region (especially in Sydney and Singapore), where they are encouraged to follow environmental guidelines to maintain water and air-quality measures. Trade and exchange is no longer happening solely through direct verbal interaction or post, but is now enabled by internet connections, social media and mobile technologies. While the extent of online interaction is increasing, the ability to create new and meaningful face-to-face encounters may be faltering (especially within the younger generation) who are more familiar with using mobile technologies for meaningful interactions.

Of the Water's Edge Public Spaces analysed, the emerging public spaces that displayed the least biodiversity are in Sydney and Hong Kong. In Singapore, the Gardens by the Bay attempts to incorporate biodiversity, but contains plastic ornamental plants inside the giant trees, this fails the original intent/ essence of the project. The giant domes are not as energy efficient as desired albeit the district cooling plants, and, the botanical and floral exhibitions involve constant re-planting to meet changing tourist expectations. All this means that emerging public spaces in Sydney, Hong Kong and Singapore are still relatively sterile regions, lacking consistent biodiversity enhancement, and this makes 'biodiversity' the one aspect where established/ historic Water's Edge Public Spaces have an advantage. The emerging public spaces show an attempt to try to connect with environmental *values*, but are not (yet) able to connect with existing environmental, biological and ecological *cycles*. This means that they are highly energy reliant spaces, and do not sustain themselves generationally. They require constant care, maintenance, planning for re-plantation and upgrading. Many of the emerging public spaces display extensive thinning to total loss of existing vegetation, replantation of single species causing fragmentation of habitats and chemical application after plantation and suppression. An evolutionary space, i.e. one that changes and grows naturally and with autonomy, or mostly in line with nature, and yet meets the needs of the users of the space, is what designers are currently aiming to create.

3. The Concepts of Green Urbanism and Bio Mimicry as Applied to WEPS

A lot has been written about green urbanism, bio mimicry and how urban design concepts may guide the inclusion of biodiversity in an urban context. Preserving biodiversity in WEPS in the face of urbanization, habitat fragmentation, environmental degradation and climate change is one of the greatest challenges of our time. In biodiversity conservation, achieving social equity along with economic and environmental benefits – the triple bottom line – is commonly seen as the key to successful outcomes. In this mix are now issues and the advantages of culture, politics and technology.

Urbanism itself can be considered a way of life that is often related to built environments. It is also considered an academic field concerned with understanding the spatial organisation and dynamics of urban areas. It relies upon comparing, analysing, studying, evaluating and inventing new ways to maintain balance in cities – between public and private, and local and global perspectives.

Green urbanism is a holistic concept of sustainable urban systems, enabling existence and change (growth or shrinkage) without negatively impacting planetary life support systems and ecosystems. It is a particular form of urbanism that is concerned with a healthy balance between the water's edge, city and its hinterland. Green urbanism also underpins practical action in order to shape the urban environment in a sustainable way (Figures 1, 2 and 3).

The concept of green urbanism arose around the late 1990s, promoting compact energy-efficient urban development, seeking to transform and re-engineer existing city districts and to regenerate the post-industrial city centre (Wheeler and Beatley, 2004; Lehmann, 2005).

Green urbanism has since been adopted in many parts of the world, and several Australian cities have formulated metropolitan plans that continue to espouse strategies such as alternative energy generation, using combined-heat-and-power (CHP) co-generation within districts, and integrating more green space to mitigate the urban heat island effect.

Other factors such as the risk of urban flooding, resource recovery through local waste recycling, efficient public transport, biodiversity protection, and the establishment and re-establishment of ecological connections have also been considered in these plans.

It must be noted, however, that in order to enable sustainable urban development and to ensure that eco-districts are successful on many levels, all urban design components and sectors need to work interactively and cannot be considered separately. This requires a holistic study of existing lifecycles associated with each development site, including the study of energy flow patterns within the site. Where these may have been broken or destroyed, this involves re-establishment of native species and their lifecycles over time.

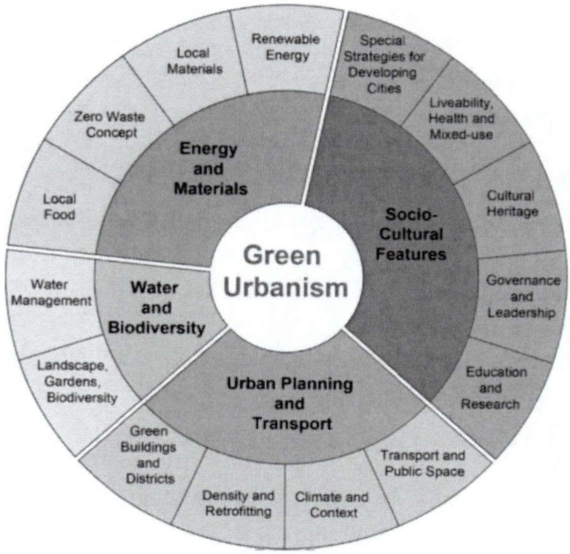

Figure 1. The green urbanism wheel shows the 15 principles grouped in four sections (Lehmann 2008).

Figure 2. The three pillars of green urbanism, and the interaction between them (Lehmann, 2010).

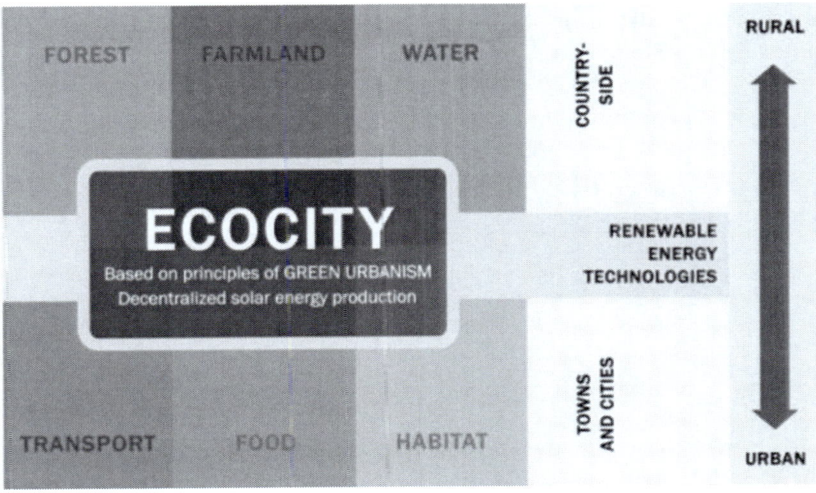

Figure 3. The holistic concept of the eco-city re-establishes a balanced, healthy relationship between the urban (city core), the rural (countryside) and the peri-urban (suburb) (Lehmann, 2008).

The tokenism associated with planting for one generation and then redoing maintenance programs emphasises existing unfortunate arrangements in the WEPS studied, where the environmental, biological and ecological cycles have been broken, and are not able to be re-established to effectively self-sustain generationally.

The principles of green urbanism are therefore based on the triple-zero framework of:

- zero fossil-fuel energy use,
- zero waste, and
- zero emissions.

Table 1 outlines the fifteen principles of green urbanism and provides sample recommendations of how they could be implemented (Lehmann, 2010).

Figures 1 and 2 show the inter-connections between the holistic principles of green urbanism while Figure 3 shows how eco-cities will enable the appropriate harnessing of various components.

Table 2 outlines the fifteen principles of green urbanism and provides sample recommendations of how they could be implemented in Water's Edge

Public Spaces in the Asia-Pacific region (including global cities like Sydney, Hong Kong and Singapore).

Table 1. The fifteen principles of Green Urbanism (Lehmann, 2010)

The 15 Principles of Green Urbanism	Sample Recommendations
PRINCIPLE 1 CLIMATE AND CONTEXT	
All urban development must be in harmony with the specific characteristics, site factors and advantages of each location. Every city is different and will need to identify its unique solutions.	Start any strategic planning process by mapping city's location and identifying its assets. Plan for development that works with the urban climate and bio-regional context.
PRINCIPLE 2 RENEWABLE ENERGY FOR ZERO CO_2 EMISSIONS	
The city as a self-sufficient on-site energy producer, using decentralized district energy systems for zero CO_2 emissions through renewable, smart electricity grids for efficient use of energy.	De-carbonize the energy supply; increase solar power to 10 per cent share in energy mix by 2020; install smart grids and make solar hot water mandatory; generate at least 50 per cent* energy on-site using renewable
	sources including precinct-scale wind turbines and small-scale biogas plants. Encourage this through feed-in tariffs.
PRINCIPLE 3 ZERO-WASTE CITY	
The zero waste city as a circular, closed-loop eco-system through waste management and resource recovery that stops waste materials from going to landfill or incineration.	Implement" zero waste city" ideas and plans; increase resource recovery rate towards 100 per cent and stop land filling. A minimum 50 per cent of on-site renewable energy generation should be the aim of all planning proposals,* with the energy mix coming from decentralized energy generation and taking into account locally available resources, as well as the cost and availability of the technology.
PRINCIPLE 4 WATER	
The city with closed urban water management and high water quality allows for sustainable water management to ensure water security through wastewater treatment and sensitive urban water management.	Power desalination by solar and recycle wastewater; keep fresh water consumption below 125 liters p. person p. day.* Construct wetlands to purify and recycle grey water.
PRINCIPLE 5 LANDSCAPING, GARDENS, GREEN ROOFS AND BIODIVERSITY	
The city that integrates landscaping, urban gardens and green roofs to maximize urban biodiversity, including landscape strategies of productive open spaces.	Continue tree planting programs and increase tree planting in streets. Constructing wetlands to purify and recycle grey water (see Principle 4 recommendation) will also improve landscaping and biodiversity.
PRINCIPLE 6 TRANSPORT AND SPACE	
The city of eco-mobility, with a good public space network and an efficient low-impact public transport system for post-fossil-fuel mobility.	Invest over 6 per cent of GDP in public transport,* expand tramlines and introduce free hybrid buses. The central city should offer multi-modal public transport systems: a high number of connections and choices,

The 15 Principles of Green Urbanism	Sample Recommendations
	frequent trams and buses, safe pedestrian and cycle networks. Improve streets by giving greater priority to pedestrians and cyclists.
PRINCIPLE 7 LOCAL MATERIALS	
The city constructed using regional, local materials and applying pre-fabricated modular systems.	Use engineered timber construction systems and make recyclability and re-use of construction elements compulsory. Use local materials with less embodied energy.
PRINCIPLE 8 DENSITY & RETROFITTING	
The city with retrofitted districts, urban infill, and densification/intensification strategies for existing neighbourhoods.	Continue street upgrading and introduction of bike lanes. Improve pedestrian connectivity with additional crossings. More useful open public space with natural elements designed for active living.
PRINCIPLE 9 GREEN BUILDINGS AND DISTRICTS	
The city that applies deep green buildings design strategies and offers solar access for all new buildings using passive design principles.	Re-introduce passive design principles for building approvals and demand higher ratings. Better adoption of energy-saving building designs and full home insulation. Offer better housing choices and more diversity in urban infill.
PRINCIPLE 10 LIVABILITY, HEALTHY COMMUNITIES AND MIXED-USE	
The city with a special concern for affordable housing, mixed-use programs, and a healthy community.	Include min. 25 per cent affordable housing in every development* and use modular prefabricated construction systems. Reduce taxation of inner-city housing, increase retrofitting and adaptive re-use of heritage buildings.
PRINCIPLE 11 LOCAL FOOD	
The city for local food supply, with high food security and urban agriculture.	Introduce urban farming in at least 20 per cent of the public parks.* Create community gardens. Maintain urban hinterland for food production. Restore connections to natural ecosystems within the urban ecosystem.
PRINCIPLE 12 CULTURAL HERITAGE	
The city of public health and cultural identity: a safe and healthy city, which is secure and just.	Protect existing structures; demand and facilitate more adaptive re-use by relaxing building code. Consult and involve communities to ensure genuine commitment.
PRINCIPLE 13 GOVERNANCE & LEADERSHIP	
The city applying best practice for urban governance and sustainable procurement methods.	Innovate public policy, roll-out incentives and deliver subsidy program for clean-tech sector. Create public-private partnerships to facilitate change, involve community groups and NGOs.

Table 1. (Continued)

The 15 Principles of Green Urbanism	Sample Recommendations
PRINCIPLE 14 EDUCATION, RESEARCH & KNOWLEDGE-SHARING	
The city with education and training for all in sustainable urban development.	Invest 3 per cent of GDP in research and innovation*; strengthen university educational programs to include climate change impact. Facilitate sustainable behaviours and create incentives for long-term behaviour change by positively influencing values and attitudes to reduced consumption. Remove policies that encourage wasteful consumption (e.g. fuel subsidies).
PRINCIPLE 15 STRATEGIES FOR CITIES IN DEVELOPING COUNTRIES	
Particular sustainability strategies for cities in developing countries, harmonizing the impacts of rapid urbanization and globalization.	Cities require adjusted strategies appropriate for the developing world, e.g. low-cost building and mass housing typologies for rapid urbanization; gender-sensitive and pro-poor approaches to development.

* All suggested figures are benchmarks derived from current best practices.

Table 2. The fifteen principles of Green Urbanism applied to Water's Edge Public Spaces (WEPS)

The 15 Principles of Green Urbanism	Sample Recommendations
PRINCIPLE 1 CLIMATE AND CONTEXT	
All urban development in WEPS must be in harmony with the specific characteristics, site factors and advantages of each location. Every site is different and will need to identify its unique strengths, weaknesses and foci.	Start any strategic planning process by mapping the Water's Edge location and identifying assets. Plan for development that works with the hydro climate, biological, ecological and environmental context. Consider specifically existing/original cyclical patterns and energy flows in the subject site.
PRINCIPLE 2 RENEWABLE ENERGY FOR ZERO CO_2 EMISSIONS	
The WEPS as a self-sufficient on-site energy producer, using (if possible) decentralized district energy networks for zero CO_2 emissions through renewable, smart electricity grids for efficient use of energy.	De-carbonize the energy supply in Water's Edge locations; increase solar power collection in public spaces to at least 3 per cent share in energy mix by 2020; install smart grids; generate at least 5-10 per cent* energy on-site in public spaces using renewable sources including precinct-scale wind turbines and small-scale biogas plants.
PRINCIPLE 3 ZERO-WASTE CITY PRINCIPLES APPLIED TO WATERS EDGE PUBLIC SPACES	
The zero waste Water's Edge Public Space as a circular, closed-loop eco-system through waste management and resource recovery that stops waste materials from going to landfill, incineration or causing pollution.	Implement "zero waste" ideas and plans into emerging public spaces; A minimum 50 per cent of on-site renewable energy generation should be the aim of all planning proposals including in Water's Edge Public Spaces,*

The 15 Principles of Green Urbanism	Sample Recommendations
This space harnesses environmental, biological and ecological cycles (including variables and parameters) effectively.	with the energy mix coming from decentralized energy generation, solar, hydro, wind or bio power, taking into account locally available resources, as well as the cost and availability of the technology.
PRINCIPLE 4 WATER	
The WEPS with closed urban water management and high water quality allows for sustainable water management to ensure water security through wastewater treatment and sensitive urban water management.	Power WEPS by energy generation options and recycle wastewater, including running desalination projects; keep fresh water consumption below 125 liters p. person p. day.* Construct wetlands in parks adjoining WEPS to purify and recycle grey water before it enters natural watercourses.
PRINCIPLE 5 LANDSCAPING, GARDENS, GREEN ROOFS AND BIODIVERSITY	
The WEPS that integrates landscaping, urban gardens and green roofs to maximize urban biodiversity, including landscape strategies of productive open spaces.	Continue tree planting programs in WEPS and increase these in key locations to act as wind and other buffers to residential areas. Constructing wetlands to purify and recycle grey water (see Principle 4 recommendation) will also improve landscaping and biodiversity.
PRINCIPLE 6 TRANSPORT AND SPACE	
The WEPS of eco-mobility, with a good network and an efficient low-impact public transport system for post-fossil-fuel mobility.	Create connections to interesting and unique WEPS, drawing tourism and city life. Invest over 6 per cent of GDP in public transport,* expand tramlines and introduce free hybrid buses. The central city should offer multi-modal public transport systems: a high number of connections and choices, frequent trams and buses, safe pedestrian and cycle networks. Improve walkability to WEPS and streets by giving greater priority to pedestrians and cyclists.
PRINCIPLE 7 LOCAL MATERIALS	
The WEPS constructed using regional, local materials and applying pre-fabricated modular systems.	Use flexible designs, environmentally triggered and floating platforms to access water, perhaps using engineered timber construction systems and make recyclability and re-use of construction elements compulsory. Use local materials with less embodied energy.
PRINCIPLE 8 DENSITY & RETROFITTING	
The WEPS with retrofitted districts, urban infill, and densification/intensification strategies for existing neighbourhoods.	Continue street upgrading and introduction of bike lanes in WEPS. Improve pedestrian connectivity with additional crossings. More useful open public space with natural elements designed for active living.
PRINCIPLE 9 GREEN BUILDINGS AND DISTRICTS	
The WEPS apply green building design strategies and offers solar access for all new buildings using passive design principles.	Re-introduce passive design principles in WEPS for building approvals and demand higher ratings. Better adoption of energy-saving building designs and full home insulation.

Table 2. (Continued)

The 15 Principles of Green Urbanism	Sample Recommendations
	Offer better housing choices next to the water and more diversity in urban infill.
PRINCIPLE 10 LIVABILITY, HEALTHY COMMUNITIES AND MIXED-USE	
The WEPS adjacent affordable housing, mixed-use programs, and a healthy community.	Diversify uses in WEPS, incorporating varied activities and flexibility of use during different hours of the day. Include min. 25 per cent affordable housing in every development* and use modular prefabricated construction systems. Increase retrofitting and adaptive re-use of heritage buildings where these exist in WE areas.
PRINCIPLE 11 LOCAL FOOD	
The WEPS with a local food supply, with high food security and urban agriculture.	Introduce urban farming in WEPS in at least 20 per cent of the public spaces.* Create community gardens beside the water. Maintain urban hinterland for food production. Restore connections to natural ecosystems within the urban ecosystem.
PRINCIPLE 12 CULTURAL HERITAGE	
The WEPS focusing on public health and cultural identity: a safe and healthy WE, which is secure and just.	Protect existing structures in WEPS; demand and facilitate more adaptive re-use by relaxing the building code. Consult and involve communities to ensure genuine commitment to this change.
PRINCIPLE 13 GOVERNANCE & LEADERSHIP	
The WEPS applying best practice for urban governance and sustainable procurement methods.	Include Innovative aspects in public policy especially for WEPS, rolling out incentives to deliver subsidy programs for clean-tech sector. Create public-private partnerships to facilitate change and development of waterfronts, involve community groups and NGOs.
PRINCIPLE 14 EDUCATION, RESEARCH & KNOWLEDGE-SHARING	
The WEPS with education and training for all in sustainable urban development.	Invest 3 per cent of GDP in research and innovation*; strengthen global university educational programs to include climate change impact and environmentally focused flexible designs. Facilitate sustainable behaviours and create incentives for long-term behaviour change by positively influencing values and attitudes to reduce consumption. Remove policies that encourage wasteful consumption (e.g. fuel subsidies).
PRINCIPLE 15 STRATEGIES FOR WEPS IN DEVELOPING COUNTRIES	
Particular sustainability strategies for WEPS in developing countries based on social, environmental and cultural context, harmonizing the impacts of rapid urbanization and globalization.	These WEPS will require unique strategies and solutions appropriate for the developing world, e.g. low-cost building and mass use areas due to rapid urbanization; these spaces should be inclusive i.e. gender, class, innovation, access and economically focused, involving sensitive approaches to development.

When Green Urbanism principles are applied to Water's Edge Public Spaces, they make the redesign, upgrade and renewal of these spaces formulaically more successful, and the composition of these spaces become balanced, lasting and rounded.

Green Urbanism and bio-mimicry have their strengths and weaknesses as concepts of sustainable urban systems, and, evaluation models to determine strengths and weaknesses based on these concepts are currently being studied. Species, their numbers and diversity are cues while studying urban waterfront diversity, and over time a stable state and re-establishment rather than reduction is sought. To enable this, more environmental, biological and ecological data is required to study biodiversity.

"Bio-mimicry is the science and art of emulating nature's best biological ideas to solve human problems"(Bio-mimicry Institute, 2009). Bio-mimicry is the technological application of designs that are already used in nature to solve problems in fields such as engineering, material science, medicine, and in other fields (Mueller, 2008).

In other words, bio-mimicry is a form of reverse engineering, whereby researchers identify a problem, seek a biological source for this problem that has evolved a natural solution, attempt to understand that natural solution, and then redesign the solution for human benefit (Gardner, 2012). While bio-mimicry assists with some problem solving for designers, it does not yet enable solutions for generational continuity. It has been argued that a humanist version of architecture and design (Kaplinsky, 2006), one that goes beyond bio-mimicry and promotes green urbanism, is necessary.

Human beings are limited to solving problems based on their ability, ingenuity, intelligence, experience and talent. While biology achieves many things that engineers may not mechanically be able to resolve, emerging biotechnology companies are trying to find solutions that go beyond bio-mimicry, into fields such as bio-engineering and nano-technology.

To harness the reproductive, replicating and autonomous nature of biological systems, architectural components should incorporate natural elements that enable replication and generational viability, i.e. allow for variables and parameters that influence seasonal and cyclical elements.

4. Biodiversity and Predator-Prey Models in the Selected Case Study Locations in Sydney, Hong Kong and Singapore

Biodiversity itself is generally defined as “The variety of life forms, the different plants, animals and microorganism, the genes they contain, and the ecosystems they form. It is usually considered at three levels: genetic diversity, species diversity and ecosystem diversity” (DECW Biodiversity Strategy, 2011). In line with this, spatially the "urban-rural gradient" provides a way to examine general patterns, usually we consider, from the water's edge to the inner city and then to hinterlands and surroundings. Changes in temperature and weather conditions based on urban built form and greenery are seen in the "physical gradients" and the "habitat-loss gradient" considers a gradient of species as habitats are lost and scatter to find replacement habitats in the urban setting (McKenny, 2002).

Applying this to the case study cities, the "urban-rural gradient" sees high biodiversity in water's edge spaces, to low level biodiversity in city cores (in Sydney, Hong Kong and Singapore), and high biodiversity as we move further away from the city. In terms of "physical gradients" temperatures are highest in urban centres (in Sydney, Hong Kong and Singapore) due to the density of built forms and lack of green spaces. Finally, the "habitat-loss gradient" is highest in most redeveloped waterfronts, followed by areas where redevelopment has happened, the gradient is lower where re-establishment of species happens in time.

This is because many development sites are cleared prior to the laying of foundations and the establishment of a building or urban space. Unfortunately, clearing of habitats affects flora and fauna adversely. Once cleared, the sites need an active re-establishment program for flora and fauna, and this recovery extends for years. Sites that remained functional for many years tend to contain greater biodiversity for this very reason, as re-establishment often takes generations, energy, effort and resources. Areas of active development usually tend to have low biodiversity due to the devastating impact on native species of most residential and commercial development methods. To re-iterate, before construction of most residential and commercial buildings, it is common for developers to remove most vegetation and even topsoil (Sharpe et al. 1986), even in prominent waterfront sites. This usually reduces construction costs by allowing equipment ready access to the construction site, but causes various related environmental issues.

If we use Blair's (2001) approach to analysing biodiversity, we can classify species in one of three categories: "urban avoiders," "urban adapters," and "urban exploiters". In doing so, we understand that while birds are the best-studied taxa for work on urban-rural gradients, these three categories have also been used in the past for work on butterflies (Blair and Launer, 1997) and lizards (Germaine and Wakeling, 2000). Given this, and the species involved in prior studies, we suggest that a study of species in terms of avoidance, adaptation and exploitation in WEPS often needs to be carried out as part of any initial architectural site analysis.

These categories show that, even in highly urbanized water's edge environments, species are non-randomly assembled in ways that approximate community assembly processes in nature. There is a social logic to the spatial distribution and movement of species. Each of these assemblages has a distinctive set of ecological characteristics that reflect the impacts of urbanization on native species.

One of the most important traits that separate the three categories is the extent to which selected species depend on human-subsidized resources to exist in water's edge and other areas (Johnston, 2001). As subsidized resources generally increase toward the urban core, there is a concurrent increase in parasitic species that utilize them. Urban exploiters are generally entirely dependent on human subsidies (i.e. obligate parasites). Urban adapters are able to utilize subsidies but they also use natural (wild-growing) resources. Urban avoiders generally tend to rely only on natural resources (Johnston, 2001). These were all investigated in the WEs analysed.

In Sydney, Hong Kong and Singapore, the emerging public spaces clearly displayed least biodiversity. The emerging spaces had all been cleared before redevelopment, and re-plantings are not yet established (especially in Hong Kong and Singapore) i.e. they are in a process of being re-established, with no greenery programs currently present in Sydney, an active gardening program in Hong Kong, and with a commercialised lack of acknowledgement of detail in Singapore. The intention of this critique is not to find flaws in existing development activity, but to remind designers of the importance of site-specific planning, consistent designing while focusing on detail, to enable and ensure existing biodiversity survives. and not to stray from the original intents of the brief when fine grained considerations are made.

In Sydney, both Circular Quay and Walsh Bay were cleared prior to project commencement, so when compared to Barangaroo, the new area for development too was cleared for development. At least, in the case of Walsh Bay, some adaptive reuse principles had been put into place. But, being more established and used, both these spaces (Circular Quay and Walsh Bay) are also "lived-in" today, and therefore more diverse in terms of the species they contain.

Table 3. Water's Edge Public Spaces analysed as early, mid and emerging lifecycles

	EARLY	MID	EMERGING
Sydney	Circular Quay	Walsh Bay	Barangaroo
Hong Kong	Star Ferry Pier	Avenue of Stars	Vic Harbour
Singapore	Boat Quay	Marina Bay Promenade	Gardens by the Bay

Figure 4. An artist's impression of Barangaroo, and the increasingly important requirement to consider issues that arise from broken natural environmental connections, loss of biodiversity and fragmentation (courtesy Lendlease, 2013).

In Hong Kong, Star Ferry Pier displays a more environmentally connected design, because birds and wildlife have established themselves in the region over time. Plants, however, are not actively present (in sufficient clusters and corridors) in this waterfront location. In Singapore, Boat Quay has the same issues in terms of plantings, and in Sydney, Circular Quay presents the same problems in terms of flora.

Combining the information in Tables 2 and 3 shows that early water's edge public spaces currently display larger species biodiversity than mid or emerging water's edge public spaces. The number of species (flora) and those protected in each space is analysed:

Sydney *Circular Quay:* High (between 1500 – 2000 species total, 6 species protected, 10kmx10km extent),

Walsh Bay: Low (Between 1000 – 500 species, 6 species protected, 10kmx10km extent)
Barangaroo: Lowest (Between 0 – 500 species, 6 species protected,10kmx10km extent)

Hong Kong *Star Ferry Pier:* High
Avenue of Stars: Low
Victoria Harbour: Lowest

Singapore *Boat Quay:* High (nearly 4000 species)
Marina Bay Promenade: Low (Between 1000-500 species)
Gardens by the Bay: Highest (over 5000)

The number of species (fauna) and those protected in each space analysed:

Sydney *Circular Quay:* High (400-500 species – 1 threatened (Epthianuraalbifrons– White-fronted Chat population in Sydney Metropolitan Catchment Management Area (Animalia Aves Meliphagidae) - 10kmx10km extent.
Walsh Bay: Low (300-400 species – 2 threatened (the above and Peramelesnasuta – Long-nosed Bandicoot population in inner western Sydney (Mammalia Peramelidae) 10kmx10km extent.
Barangaroo: Lowest (200-300 species – 2 threatened (the above) 10kmx10km extent.

Hong Kong *Star Ferry Pier:* High
Avenue of Stars: Low
Vic Harbour: Lowest

Singapore *Boat Quay:* High
Marina Bay Promenade: Low
Gardens by the Bay: Highest

Table 4. Overall biodiversity in selected case study locations

	EARLY	*MID*	*EMERGING*
Sydney	Mid biodiversity	Low biodiversity	Low biodiversity
Hong Kong	Mid/High biodiversity	Low biodiversity	Low biodiversity
Singapore	Mid/High biodiversity	Mid/High biodiversity	High biodiversity

Predator-prey models generally show two specific species' numbers over time, providing an understanding of the relationships between two or more species that are directly linked to each other. These models are often used in environmental science, but less so in architecture and design. An understanding of these models along with other environmental factors in design, enables architects, planners and urban designers to influence emerging public spaces, so that new spaces are adequately prepared for the habitats that are being affected. The Lotka-Volterra Model is a commonly-used mathematical model (see Figures 5 and 6), which often makes assumptions such as: 1) the populations in the case studies are large enough that it makes sense to treat the abundances as continuous rather than discrete variables; 2) the populations in the case study locations are well mixed in the environment; 3) the population dynamics are completely determinable, i.e. no random events are considered (this is usually where architectural methods and construction would fall); 4) the populations are closed, i.e. immigration or emigration of species does not occur; 5) in the absence of predators, prey numbers grow exponentially; 6) the per predator rate of prey consumption is a linear function of prey abundance; 7) predators are specialists and without the prey their populations will decline exponentially; 8) the rate with which consumed prey are converted to new predators is a linear function of prey abundance; 9) both populations are unstructured (age, sex, size wise etc.); and 10) reproduction follows feeding and the process continues in a practical mathematical sense, models based on these are assumptions, showing predator-prey numbers over time are then generated (Krivan, 2008).

The relationship between existing species is complicated, and not often only a predator-prey relationship, and this work has been conducted merely to consider the changing state of a species' environment, i.e. if any change has occurred to 'the designing of space' that involves site analysis, from an environmental, biological and ecological context.

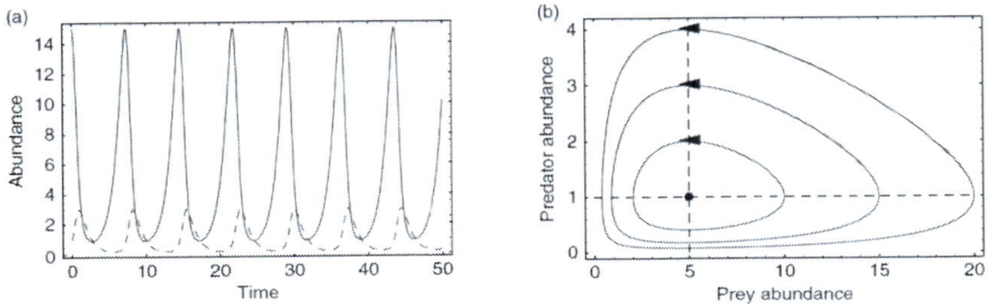

Figure 5. The solutions of the Lotka-Volterra model showing predator and prey abundance and their inter-relationships, and extensions (Krivan, 2008).

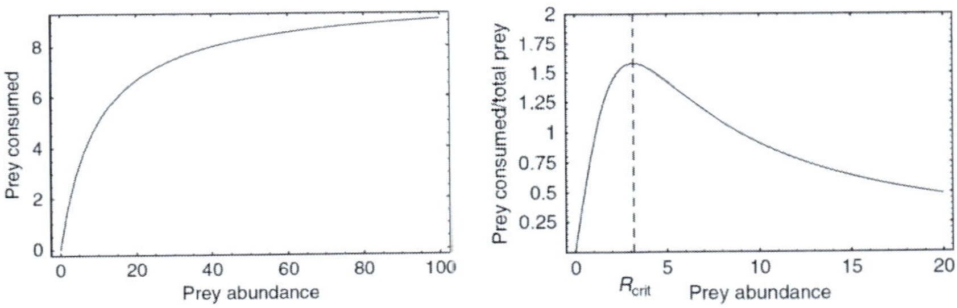

Figure 6. Functional response to analysis. the different options within systems over time (Krivan, 2008).

Human beings naturally eliminate pests, making environments safe for habitation and recreation, but in affecting the environment, for example site clearance for development, not only is one species is affected, its prey is also affected, and this change travels down the food chain affecting not only that chain, but also inter-reliant and inter-dependent chains, and associated lifecycles in the future. It is therefore important to maintain a certain ecological balance, for the benefit of the overall environment. The challenge currently faced by environmentally concerned designers is to make an accurate determination about species preservation and lifestyle in order to maintain biodiversity.

Various events cause mortality to predators or their prey or both. The effect of mortality events on predator–prey dynamics will depend on the event itself. An extreme example would be where a large fire kills off a predator species, along with its prey in a particular locality, and in such instances, re-establishment of the prey may occur before the resurgence of the predator (Terry, 2013).

The available data in Sydney, Hong Kong and Singapore did not match or make possible the detailed study of predator-prey models and its extensions, but the understanding of these models can facilitate better design processes, as practitioners become aware of the issues caused by loss of native flora and fauna, fragmentation and destruction of ecological and biological balances and connections. Much remains to be attained in our emerging public spaces in terms of ecological and environmental balances and connections, biodiversity, autonomy, energy generation and generational viability.

5. Other Key Australian Examples of WEPS

Within Australian urban, suburban, regional and rural contexts, numerous WEPS sites begin construction with demolition, including a removal of all flora and fauna contained in the site, breaking all environmental, biological and ecological connections, before development commences and then a slow rehabilitation takes place. On the water's edge, such a process leaves environmental connections faltering, taking longer than usual to reconnect because of the sensitive nature of the system, the consequences of breaking cycles and associated seasonal interdependencies. This can be seen in places such as Elizabeth Quay in Perth, Mildura in Victoria next to residential developments beside the Murray River, and in Port Adelaide where there has not been adequate attention to the details of the urban design.

These areas now display opportunities for appropriate environmental rehabilitation and the reinstatement of connection.

Photograph 1. Elizabeth Quay Waterfront development, Perth, WA.

Photographs 2, 3 and 4. Emerging WEPS in Murray Bridge, Hume Reserve, including the ARTC owned Railway precinct and historic riverfront.

Photographs 5, 6 and 7. Existing WEPS in Wentworth, where the Murray River and the Darling River meet.

Photographs 8, 9 and 10. Emerging WEPS in Mildura, where residential development fronts a cafe and house boats.

The city and regional/rural Australian examples chosen are from Perth, Brisbane and Sydney, and, Mildura, Renmark and Wentworth respectively. Local examples display a need for care, upgrading and financial assistance. The community supports these upgrades but funding is often the most crucial issue when development is planned. It is essential that economic development activities on the waterfront bring resources to regional areas, even though the downside is the potential to reduce quality in the public realm for everybody. This is noticeable in some designs, surely unanticipated by the developer, where for example, some sections of Mildura's public water's edges (by the river) are difficult to walk along because of bird droppings covering the public walking path. However, this is counteracted in developed scenarios where the birds pick up food scraps in cafes.

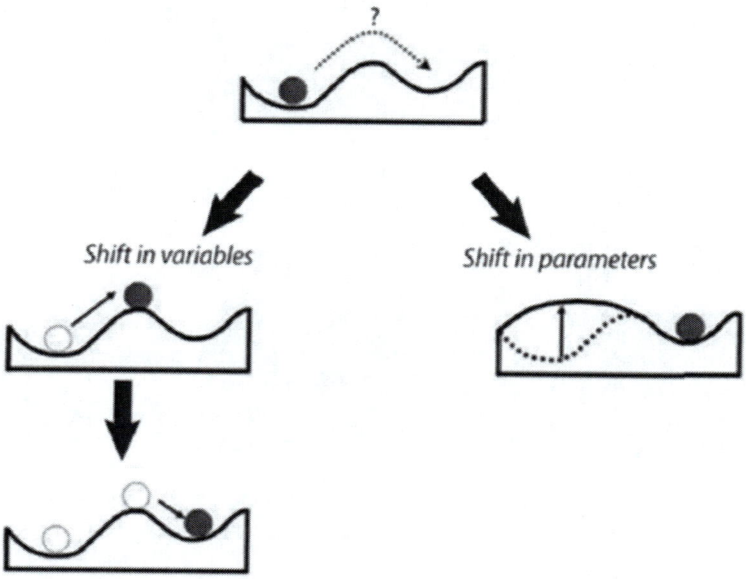

Figure 7. Causes of change and consideration of these in architectural science. Adapted from "Alternative stable states in ecology" (Baisner et al., 2003) and different types of stability in theoretical communities (Ives and Carpenter, 2007).

6. Ecological Stability as a Natural Process?

Ecological stability is not a natural process, and cycles and environments often change, affecting population numbers. Amidst this change, human activity has arguably been identified as the primary cause for environmental degradation and climate change. Whether or not this is the case, conservation scientists often aim for stable states or manageable population rates. A change in population stability can generally be linked to resource loss i.e. affecting the population of prey, and/or other naturally occurring mineral or plantations that assist the species to survive.

As such, stable states within the system seem fleeting, but population numbers fluctuate in these states even seasonally, keeping overall numbers static or similar over time (i.e. incorporating births, rebirth and death). In marine protection areas, or areas where heritage agreements exist to protect native vegetation (e.g. Monarto, SA), ecosystem models may be put into place and managed (Brochier, 2013).

Systemic change occurs due to a shift in variables or a shift in parameters (Figure 7). The variables may be enabled due to catalysts, or the parameters may be altered due to seasonal cycles (John, Lehmann and Sivam, 2013).

7. The Ecological Balance – Connection and Re-Establishment of Diversity through Design and after Construction – Understanding Native Plants and Ecological Succession

Preserving ecological balances means understanding constantly changing systems, and in practice, this is commonly executed through trend analysis, future forecasting and mapping but more immediately through preservation, bioremediation and appropriate designing. This can be further understood using predator-prey studies, including through acceptance of systems thinking as in 'the game of life', 'Daisy world' and other such models.

According to McKenny (2002), commonly known replacement types of habitats have an array of increasing habitability to most native species, and decreasing proportion of coverage toward the urban core. This did not consider human impact on Water's edges over time, as places where exploration commenced, and as areas that are either highly bio-diverse, or currently being rehabilitated. The types of land based habitats are the:

1. **Built habitat:** buildings and sealed surfaces, such as roads
2. **Managed vegetation:** residential, commercial, and other regularly maintained green spaces

3. **Ruderal vegetation:** empty lots, abandoned farmland, and other green space that is cleared but not managed
4. **Natural remnant vegetation:** remaining islands of original vegetation (usually subject to substantial normative plant invasion)

Additionally, a new type should be added, i.e.

5. **Water's Edge:** adjacent to the water, either ocean-side, seaside, riverside or lakeside.

In their book 'Tropical Conservation Biology', Sodhi et al. argue that it is important to realise that extinctions are a normal part of evolution, but human modifications can accelerate the rate of species lost, at local to regional to global scales. Habitat loss is indeed the main driver of tropical extinctions but it may act synergistically with other drivers such as overharvesting, alien species invasion and climate change. Extinctions are assumed to disrupt vital ecological processes, leading to ecosystem collapse (Sodhi, Brook and Bradshaw, 2007).

Public biodiversity education would be most effective in terms of conservation and prevention, but only if we draw on knowledge of species to promote an understanding of concepts such as ecological succession and the role of native plants in promoting native animal diversity. This knowledge can be dispersed through traditional and online methods of teaching and information transfer.

8. Considerate Designing for WEPS

In 2005, Balmford and Bond highlighted four areas where better data is currently required: (1) measuring rates of habitat and population changes, (2) allowing predictions of the future state of biodiversity and natural ecosystems, (3) understanding the divers of biodiversity and habitat change; and (4) understanding the link between nature, ecosystem services and human wellbeing. This is true for WEPS in the Asia-Pacific region, where better data gathering is recommended; however, based on available data, the authors believe that considerate designing incorporates the following steps:

1. Completing an in-depth site analysis that includes collection of species data present in existing ecological models,

2. Understanding the environmental, biological and ecological systems that can be harnessed (considering variables, parameters and catalysts that caused change),
3. Designing to incorporate holistic green urbanism, humanistic, bio-mimicry related integrated multi-disciplinary principles,
4. Incorporating low-carbon, energy efficiency focused and zero waste principles, and,
5. Allowing designs that are autonomous, flexible and change over time with the changing environmental, biological and ecological systems.

9. Changing the Design Process –'Inclusive Interdisciplinary Thinking' for WEPS

Interdisciplinary thinking allows various different systems to work together, making the most of apparently unconnected but linked aspects of the ecology, environment, biology and other fields. Cabrera (2006) found that in contrast to critical thinking, systems thinking is interdisciplinary and may also act as a bridge between the physical, natural, and social and other sciences (including architectural sciences). The authors believe that there are various specialists working on models and processes that could be better integrated. The thought processes in these individual projects are often critical, aiming for highest potential etc., but an integrated systems thinking approach can often be more inclusive and a better governance and engagement fuelled approach. It is interesting to note that 88% of the papers in which the term 'critical thinking' appeared are still in the social sciences, arts, and humanities literature, whereas ‘systems thinking’ appeared in only 48% of the literature from those fields, with the remainder dispersed across the disciplinary spectrum from business, administration, finance, and economics, to engineering, computer science, and mathematics, to physics, astronomy, and planetary science (Cabrera et al., 2008). This suggests that certain disciplines could adopt a more inclusive interdisciplinary thinking model. This interdisciplinary dispersal is increased when systems’ thinking is combined with its near relation “systems science”, and we might also consider systems thinking’s applicability to "architectural science". Here, and in associated fields, systems thinking may also act as a bridge between the academic, professional and lay communities, providing feedback between “what we know about systems” (e.g., systems sciences) and “the conceptual patterns of how we think systemically” (e.g., systems thinking) (Cabrera, 2006), and applying these effectively in an integrated architectural/practical manner.

As one ventures into the complexities of systems thinking literature, it is important to realize and to remember that systems thinking could become increasingly popular because it provides a new way to think about, or conceptualize the world around us, whether our issues rest within a local or global context. Combining varied disciplines and using an inclusive approach will reap overall benefits. Interestingly, because the construction of systems thinking is unclear, people who view systems thinking as a kind of solution see its potential even when they do not entirely understand what it is all about. The authors suggest that there are benefits in using many evaluators and their clients, funding agencies, program planners, field staff, and other stakeholders, and to have their consistent involvement in the evaluation process.

There are varied and integrated ways to think about the basic constructs of systems thinking. Some scholars and evaluation practitioners view this as a specific methodology, similar to 'system dynamics', while others believe it is a "plurality of methods" (Williams & Imam, 2006). Some see systems thinking as 'systems science', while others see it as a general systems "theory". Still others see systems thinking as a social movement or new interdisciplinary inclusive way of thinking. The authors propose that systems' thinking is both *conceptual and useful*, because changing the way we think involves changing the way we conceptualize and make use of information architecturally. That is, while systems' thinking is informed by systems ideas, systems methods, systems theories, the systems sciences, and the systems movement, it is, in the end, different from each of these, and its usefulness practically and theoretically differs with the different uses of it.

In short, *systems thinking* can be seen as a formal, abstract, inclusive and structured cognitive endeavour. While not all systems are complex, all thinking is intricate and complex, and as such, the process of thinking in a systemic way can be a training process but is inevitably complex. "Systems thinking is also based on contextual patterns of organization rather than specific content" (Cabrera, 2008). For example, systems thinking may balance the focus between the whole and its parts, and takes multiple perspectives from different disciplines and contexts into account. Nobel laureate Richard Feynman (2006), as also previously referenced by Cabrera, provides a famous example of the kind of contextual patterns to which we are referring, it makes no difference that Feynman refers to specific content domains (i.e., chemistry, climatology, physics, cognition, etc.), what makes this famous quote an example of systems thinking is the way he transforms contextual patterns: he transgresses parts and wholes, takes new perspectives, forms new relationships, and makes new distinctions from a simple analogy:

> "A poet once said, 'The whole universe is in a glass of wine.' We will probably never know in what sense he said that, for poets do not write to be understood. But it is true that if we look at a glass of wine closely enough we see the entire universe. There are the things of physics: the twisting liquid which evaporates depending on the wind and weather, the reflections in the glass, and our imagination adds the atoms. The glass is a distillation of the earth's rocks, and in its composition we see the secrets of the universe's age, and the evolution of the stars. What strange arrays of chemicals are in the wine? How did they come to be? There are the ferments, the enzymes, the substrates, and the products. There in wine is found the great generalization: all life is fermentation. Nobody can discover the chemistry of wine without discovering the cause of much disease. How vivid is the claret, pressing its existence into the consciousness that watches it! If in our small minds, for some convenience, we divide this glass of wine, this universe, into parts—physics, biology, geology, astronomy, psychology, and so on—remember that nature does not know it! So let us put it all back together, not forgetting ultimately what it is for. Let us give one more final pleasure: drink it and forget it all!"(Cabrera, 2008)

One could argue that a more systemic approach (like Feynman's approach to wine taken from Cabrera (2008)) to evaluating any program would include: defining what the program is and is not; identifying the unique components (or parts) of the program; and recognizing the relationships among the parts and between each part and the program as a whole. Note that each component of the program affects the delivery of content and achievement of outcomes as a whole. Further, many evaluators who advocate approaches that include multiple stakeholders in the evaluation process (e.g., participatory action research and public surveying) do so because they recognize both the importance of taking multiple perspectives to better inform the evaluation of design and to ensure that an evaluator has a comprehensive understanding of the program relative to all the people who comprise part of the system in which the program lives (Cabrera, 2008).

Ultimately, we would argue that, much like Feynman's glass of wine, any evaluation can and should be viewed in the same way that transforms contextual patterns: as parts, wholes, and the relationships among them; as well as the relationships between the program and the larger, external forces with which it rests; distinctions must be made to set boundaries on the program's scope and thus, establish criteria as to what can be measured to make assessments; and finally, the ability to take varied perspectives enables evaluators to better understand the richness of both a program's content and the system of which it is a part. Separating developed and developing contexts and applying different design solutions is therefore not an effective methodology in a changing world, because ultimately, each system has varied members and changing relationships, external forces etc.

Now, let us imagine a similar example of 'inclusive interdisciplinary' thinking. The authors propose that the ideas of inclusive interdisciplinarity lead systems thinking further, pushing past the established boundaries, and making the example cited above work better. Let's look at transformation through unique blends, still transgressing parts or wholes, taking a fresh new outlook, forming intricate relationships and distinctions and building on Feynman's metaphor.

The authors believe, 'The universe can be understood if one understands the perfection of a blend of juices', like mindedly searching for the true way of getting outcomes in a team. The juices may be orange, guava, apple, carrot, celery, pineapple etc. and the coordinators and/or supervisors may be the ice, enabling the many tastes to cool together to attain the right combination of flavors. And, it is still true that if we see this glass of juice closely, it contains metaphorically, the entire universe, the things of physics: the twisting liquid which still evaporates depending on the wind and weather, the reflections in the glass, and our imagination adds the atoms. The glass is still a distillation of the earth's rocks, and in its composition we still see the secrets of the universe's age, and the evolution of the stars. What strange arrays of vitamins are in the juice? How did they come to be? There are the minerals, the fruity aspects, the substrates, and the products. Here in this juice is found the great generalization: all life is a fine blend of varied tastes and combinations. Nobody can discover the chemistry of a well-mixed fruit juice without discovering the cause of much disease. How vivid and unique is the taste, pressing its existence into the consciousness that watches it! Fresh and refreshing at the same time! If in our small minds, for some convenience, we divide this glass of juice, this universe, into parts—physics, biology, geology, astronomy, psychology, and so on—remember that nature does not know it! So let us put it all back together, not forgetting ultimately what it is for. Let us give one more final pleasure: taste its uniqueness and enjoy it all! And, be healthier tomorrow for it!

Ultimately, we argue that, much like Feynman's glass of wine and the abovementioned comparison of it with a glass of fruit juice, any evaluation can and should still be viewed in the same way that transforms contextual patterns:

as parts, wholes, and the relationships among them; as well as the relationships between the program and the larger, external forces within which it rests; distinctions must be made to set boundaries on the program's scope and thus, establish criteria as to what can be measured to make assessments; the ability to take varied perspectives enables evaluators to better understand the richness of both a program's content and the system of which it is a part. But, unlike a glass of wine, the juice is interdisciplinary in nature, combining many different types, and yet when tasted it seems it is just of one blend due to the inclusivity we entrust the different components with (e.g. the pineapple contributes to taste just as does the orange).

10. The Elements of Water's Edge Public Spaces – A Focus on Environmental Connections

This leads us back to practicality, the diagram below shows the elements of emerging Water's Edge Public Space where Incomplete Urbanism (Lim, 2012) and green urbanism and other concepts converge in association with plans for biodiversity.

Ecological restoration aims to initiate, facilitate or accelerate the recovery of an ecosystem that has been damaged; it means renewing and restoring degraded or destroyed ecosystems and habitats in the environment by active human intervention and action, for example through constructed wetlands, reforestation and re-vegetation (Clewell and Aronson, 2013).

Ecological connections may be created and reinstated, on a smaller scale within urban environments, for example in the form of urban farming, roof-top gardening and green walls. On a larger scale, all public spaces within urban areas may establish different and unique, linked or distinct ecological connections.

Forming and re-instating ecological linkages in design and architecture involves understanding the natural and seasonal cycles, to harness environmental, ecological and biological change. Appropriate architectural forms following these principles do not just mimic nature but go beyond, to ride and benefit from natural cycles (births, deaths, etc.), including catalysts that enable continual restorative change and recovery over time.

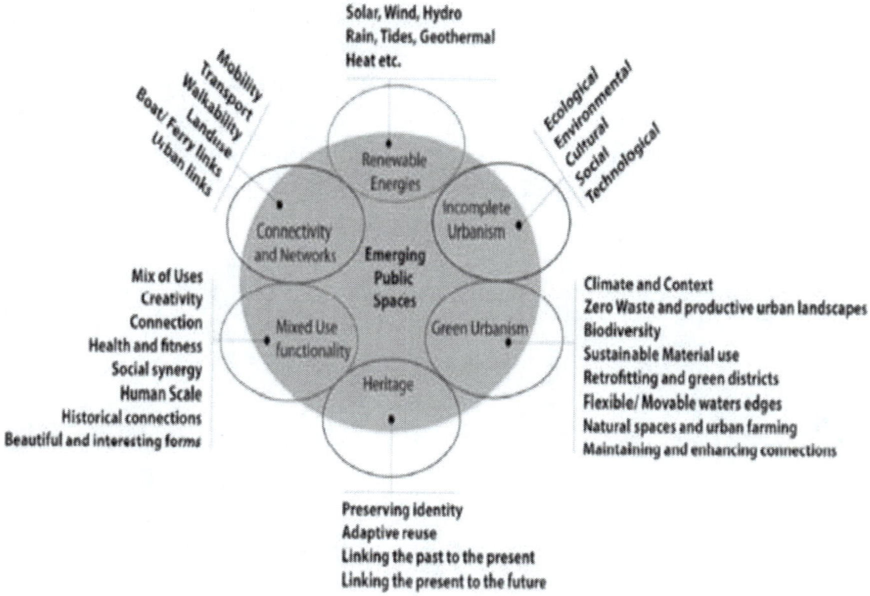

Figure 8. The elements of emerging water's edge public spaces (John, Lehmann and Sivam, 2013).

The use of appropriate materials in design can allow for the preservation of ecological balances, cycles and uses, but the design process must be flexible enough to allow the countering of ecological losses incurred through development. We suggest that effective design includes consideration firstly of individual interaction with the environment; secondly, socioeconomic and environmental factors in design; thirdly, maintaining interdisciplinary connectivity through design (green urbanism concepts and others, including harnessing solar, wind and environmental synergies); and fourthly, variables (catalysts) and changing parameters (cycles, seasons) – all this in the context of set locally specific indicators for sustainable design that are informed by interdisciplinary systemic and inclusive models.

To be truly successful, the green urbanism and zero waste city concepts must be supported by architectural designs that continually form, support and enhance ecological connections. 'Optimal ecological connections' are a product of considered design (forming, supporting and enhancing ecological links), and the desired evolution of the space through time.

11. The Variables and Parameters in Designing to Maintain and Reinstate Biodiversity

The future of our societies is not just a technical matter of finding more eco-friendly energy solutions, but a question of holistic socio-ecological change. Humans have the capacity to rapidly adapt to changing ecological conditions, perhaps more so than many animals and plants, but we have both an ethical responsibility and ecological imperative to embrace truly green urbanism, and/or sustainable systems. Some variables and parameters in designing to maintain and reinstate biodiversity are provided below:

Variables - Soil conditions, climate, temperature, replanted species, connections with existing ecological, biological and environmental cycles and processes, incorporation of a person or thing to act as a catalyst, building materials, avenues of incorporation of green spaces and other flexible, material or interest based aspects into designs.

Parameters - Seasons, biology, genetic compositions, reinstatement ability, etc.

Conclusion and Further Research

Species richness of many taxa generally declines along a gradient, starting with the highest richness along the water's edge, descending to the lowest richness (found in the urban core) and then a high richness in areas further away from the city. Urban planners, architects and designers should find ways to preserve biodiversity as cities expand outward, and subsequently modify natural habitats. Such efforts would most likely focus on preserving as much remnant natural habitat as possible, as opposed to most current land development techniques, which remove most natural vegetation before construction.

This research plans to advance planning theory in the management of contemporary urban change and support more informed planning policy in the area of Water's Edge Public Spaces. The application of green urbanism and bio-mimicry can assist with these endeavours if with humanistic and re-generation focused methodology. The authors believe that government agencies and academic institutions should work closely to allow systems focused integrated discussions related to biodiversity, and to work together during policy formulation.

REFERENCES

AFCD, HK. (2012). The *Strategic Plan for Biodiversity 2011-2020 and Aichi Biodiversity Targets*, Hong Kong.

AFCD, HK. (2013). *The Hong Kong Biodiversity Strategy and Action Plan*, Hong Kong.

Balmford, A. & Bond, W. (2005)."Trends in the state of nature and their implications for human well-being". *Ecology Letters*, *8*, 1218-1234.

Beisner, B. E., Haydon, D. T. & Cuddington, K. (2003). "Alternative stable states in ecology". *Frontiers in Ecology and the Environment*, *1*(7), 376-382.http://dx.doi.org/10.1890/1540-9295(2003)001[0376:ASSIE]2.0.CO;2

Biomimicry Institute (2009). http://biomimicry.net accessed on 13 Nov 2013.

Blair, RB. (2001). "Birds and butterflies along urban gradients in two eco-regions of the U.S". 33-56 in Lockwood JL, McKinney ML, eds. *Biotic Homogenization*, Norwell (MA): Kluwer.

Brochier, T, et al. (2013). "A multi-agent ecosystem model for studying changes in tropical estuarine fish assemblage within a marine protected area", *Aquatic Living Resources*, 147–158.

Cabrera, D. (2006). *Systems thinking.* Thesis, Cornell University, Ithaca, NY.

Cabrera, D. et al. (2008). "Systems Thinking". *Evaluation and Program Planning.*, *31*(3), 299-310.

Clewell, A. F. & Aronson, J. (2013). "Relationship of Restoration to Related Fields", *The Science and Practice of Ecological Restoration*, 199-216.

DECW NSW (2011).*Draft New South Wales Biodiversity Strategy*, Department of Environment, Climate Change and Water NSW, Sydney.

DECC (2008a). *A new Biodiversity Strategy for New South Wales*, Discussion paper, DECC 2008/398, Department of Environment and Climate Change NSW, Sydney.

Feynman, R. (2006). *The Feynman lectures on physics [UNABRIDGED]*. Audio CD. Basic Books.

Gardner, GE.(2012). "Using Biomimicry to Engage Students in a Design-Based Learning Activity", *The American Biology Teacher*, *74*(3),182-184.

Germaine, S. S. & Wakeling, B. E.(2001)."Lizard species distributors and habitat occupation along an urban gradient in Tucson", Arizona, USA. *Biological Conservation*,*97*, 229-237.

Ives, A. R. & Carpenter, S. R. (2007)."Stability and diversity of ecosystems", *Science*, *317*, 58-62.http://dx.doi.org/10.1126/science.1133258.

John, M., Lehmann, S.& Sivam, A.(2013)."The Sustainable Design and Renewal of Water's Edge Public Spaces in the Asia-Pacific Region: Sydney, Hong Kong and Singapore", *Journal of Sustainable Development*, *6*(8), 27-51.

Johnston, R. E.(2001). "Synanthropic birds of North America", pp 49-67 in Marzluff JM, Bowman R, Donnelly R, eds. *Avian Ecology in an Urbanizing World*. Norwell (MA): Kluwer.

Lehmann, S. (2005). "Towards a sustainable city centre: Integrating ecologically sustainable development principles into urban renewal", *Journal of Green Building*, *1*(3), 83–104.

Lehmann, S. (2008). "Cities in transition: new models for urban growth and neighbourhoods", in P. Droege (Ed.) *Urban energy transition: from fossil fuels to renewable power*, Elsevier, Oxford, 409-430.

Lehmann, S.(2010).*The Principles of Green Urbanism: Transforming the City for Sustainability* (London, Earthscan).

Lim, W. S. W. (2012).*Incomplete Urbanism: A critical Urban Strategy for Emerging Economies*, World Scientific Publishing, Singapore.

Kaplinsky, J.(2006). "Biomimicry Versus Humanism", *Architectural Design*, *76*(1), 66-71.

Krivan, V.(2008)."Prey-Predator Models", *Encyclopaedia of Ecology*, 2929-2940.

McKenny, M. L.(2002). "Urbanisation, biodiversity and conservation", *Bioscience*, *52* (10), 883-890.

Mueller, T. (2008)."Biomimetics: design by nature". *National Geographic Magazine*, April. Available at http://ngm.nationalgeographic.com/2008/OH/biomimetics/tom-mueller-text.html.

NPB (2010)."4th National Report to the Convention on Biological Diversity", National Parks and National Parks Board, Singapore.

Sala, O. E., Chapin, F. S., Armesto, J. J., Berlow, E., Bloomfield, J., Dirzo, R., Huber Sanwald, E., Huenneke, L. F., Jackson, R. B., Kinzig, A., Leemans, R., Lodge, D. M., Mooney, H. A., Oesterheld, M., Poff, N. L., Sykes, M. T., Walker, B. H., Walker, M.& Wall, D. H.(2000). "Global biodiversity scenarios for the year 2100", *Science*,*287*, 1770 – 1774.

Sharpe, D. M., Stearns, F., Leitner, L. A. & Dorney, J. R. (1986)."Fate of natural vegetation during urban development of rural landscapes in south-eastern Wisconsin". *Urban Ecology*,*9*, 267-287.

Sodhi, N., Brook, B. W.& Bradshaw, C. J. A.(2007).*Tropical Conservation Biology*, Blackwell Publishing, USA, UK and Australia.

Terry, A. J. (2013). "Prey resurgence from mortality events in predator–prey models", *Nonlinear analysis: Real World Applications*, *14*(6), 2180–2203.

UN ESCAP/Lehmann, S. & Thornton, K.(2013).*Planning Principles for Sustainable and Green Cities in the Asia-Pacific Region: A New Platform for Engagement*. UN ESCAP, Bangkok.

Urban Redevelopment Authority, 2008, *Master Plan 2008.* Singapore.

Wheeler, S.M. & Beatley, T. (eds.) (2004).*The Sustainable Urban Development Reader*, London: Routledge.

Williams, B. & Imam, I. (Eds.) (2006). Systems concepts in evaluation: An expert anthology, Edge Press, Point Reyes, CA.

In: Urbanization
Editor: Harold J. Kauffman
ISBN: 978-1-63117-063-8

Chapter 3

A MATHEMATICAL-MODELING APPROACH TO URBANIZATION CAUSED BY MIGRATION

Minoru Tabata[1*], Nobuoki Eshima[2] and Ichiro Takagi[3]
[1]Department of Mathematical Sciences,
Osaka Prefecture University, Sakai, Osaka, Japan
[2]Department of Statistics
Oita University, Oita, Japan
[3]Department of Business Management, School of Business Studies,
Tokai University, Kumamoto, Japan

ABSTRACT

Urbanization is caused by migration. However, an essential difference exists between migration in the past and that in the present. Hotelling constructed a mathematical model which can accurately describe the former case. In quantitative sociodynamics, Weidlich and Haag constructed a mathematical model which can accurately describe the latter case. The behaviors of these mathematical models are very different from each other, and the behavioral difference expresses the essential difference between migration in the past and that in the present. By comparing the mathematical models, in the present research we prove

* Email: mnrtabata@luck.ocn.ne.jp , Fax: +81 72 254 9916.

that the essential difference is caused by the difference between the cost of movement in the past and that in the present.

Keywords: Migration; sociodynamics; Hotelling's migration model; diffusion equation; master equation

1. Introduction

Migration plays a very important role in urbanization, and there have been a great number of studies on this subject (see, e.g., Åndersson and Philipov (1982), Beckmann and Puu (1990), Haag and Weidlich (1986), Helbing (1995), Hotelling (1978), Richard (1996), Weidlich (2000), Weidlich and Haag (1983), Weidlich and Haag (1988), Zhang (1989), and papers cited in their references). Migration has a very great influence also on human history, and history abounds with examples (see, e.g., Richard (1996)). Comparing migration in ancient or medieval times with that in the present, we find an essential difference between them. For example, in the present human beings can migrate to distant places very quickly in virtue of automobiles, trains and aircraft, but they could not do so in the past. In the present research we will take a mathematical-modeling approach to this essential difference.

First we will consider migration that caused urbanization in the past. History abounds with examples of such phenomena (see, e.g., Richard (1996)). The Greek city-states sent the colonists to less densely populated regions, as they grew stronger and larger. In the 4th, 5th, and 6th centuries, the ancient German tribes migrated westward and southward in the European Continent, and the movement is regarded as one of the main reasons for the disruption of the Roman Empire. In the 17th century a large number of people began to migrate from the European Continent to the North American Continent, where they migrated westward. This westward movement provided the foundation for the USA (see Hotelling (1978, p. 1225)). It follows from *the law of diminishing returns* that production can be carried on more efficiently in thinly populated than in densely populated places. In particular, the rise of land rent with increasing population density can supply a motive for migration (see Hotelling (1978, p. 1229)). Moreover the spirit of adventure is common to all humankind. Combining these considerations and the above examples, we deduce that human beings have a marked tendency to move from places of higher population density to places of lower population density motivated by economic potential and the spirit of adventure, which caused urbanization (see

Hotelling (1978, p. 1225)). However, it is well known in chemistry and biology that macroscopic movement of particles (e.g., cells, bacteria, chemicals, and so on) can be described as a *diffusion process* (see Stroock and Varadhan (1997) for the mathematical details of diffusion processes), i.e., that particles have a tendency to go from places where they are plentiful to places where they are scarce. Hence we see that migration is very similar to diffusion of particles in chemistry and biology. However, diffusion of particles can be described by a linear partial differential equation called *the linear diffusion equation* (the equation is one of the most fundamental equations in chemistry and biology). Based on these ideas, Hotelling in 1921 constructed a mathematical model of migration which is described by the linear diffusion equation (see Hotelling (1978, pp. 1229-1234, (20)) and Beckmann and Puu (1990, pp. 60-63, (3))). We refer to this model as *Hotelling's model.*

It is assumed in Hotelling (1978, p. 1229) and Beckmann and Puu (1990, p. 60) that each individual can migrate only to areas adjacent to a place where she/he lives. In fact, the equalities (17-18) of Hotelling (1978, p. 1229) and the equality (1) of Beckmann and Puu (1990, p. 60) cannot be obtained without this assumption. Moreover, it is deduced in Tabata and Eshima (1999, pp. 171-172) that if migration is described by the linear diffusion equation, then each individual migrates continuously in space and no individual can leap quickly to distant places. Hence, we see that each individual can migrate only to areas adjacent to a place where they live in a world where migration can be described by Hotelling's model. For example, we may consider that automobile, train, or aircraft are not available in such a situation. Furthermore, as already mentioned above, Hotelling's model is induced from examples of migration in ancient or medieval times, and in particular it is proved in Hotelling (1987, pp. 1235-1239) that Hotelling's model fits statistical data on the westward movement in the USA. Hence there is evidence that Hotelling's model can accurately describe migration in the past.

Next we will consider migration that causes urbanization in the present. Large free economic unions have been established. For example, since the 1990's many free trade areas have been established and planned such as CEFTA ('93), AFTA ('93), NAFTA ('94), SADC ('94), ANCOM ('95), and FTAA (2005). In such free trade unions, goods and services are traded freely, but migration is restricted at a certain level. Now there is a new movement to completely remove these restrictions. For example, the EU has been moving toward an economic integration of 27 countries, and it is trying to remove the restriction in the near future. If these restrictions are abolished, then in order to achieve a higher income an individual could relocate. Urbanization will be accelerated greatly by such migration, and in turn, the resulting economic potential enhanced by the urbanization will produce a large effect upon the rate of population mobility. Investigating such a phenomenon, we see that human beings have a marked tendency to migrate from places where economic potential is low to places where economic potential is high in the present. It is known in statistical physics that if a physical or chemical quantity (e.g., gas particles, plasma, chemicals, and so on) is distributed and changes stochastically, then the density of quantity can be described by a nonlinear integro-partial differential equation called *the master equation* (the equation is one of the most fundamental equations in statistical physics). Combining these considerations and the idea that stochastic phenomena in human society can be described by mathematical models in statistical physics (the idea plays an essential role in *synergetics*), Weidlich and Haag construct a mathematical model of migration that is described by the master equation (see, e.g., Helbing (1995), Weidlich (2000), and Weidlich and Haag (1983, 1988)). We refer to the model as *the master-equation model*.

In constructing the master-equation model, it is assumed that each individual is not spatially restricted to adjacent areas. Furthermore, it is deduced in Tabata and Eshima (1999) that if migration is described by the master equation, then each individual can migrate to distant places easily. From these results we see that each individual is less restricted by distance in a world where migration is described by the master-equation model. Hence, for example, we may consider that efficient transit exists (for example, automobile, train, and aircraft). Moreover, as already mentioned, the master-equation model is based upon observations on migration in the present. In fact the master-equation model fits statistical data on migration in the present (see Weidlich and Haag (1988)). Hence we deduce that the master-equation model can accurately describe migration in the present.

We must note that Hotelling's model and the master-equation model are described by two different kinds of functional equations respectively, i.e., by a linear partial differential equation and a nonlinear integro-partial differential equation respectively. In fact, the behaviors of the models are different from each other in general. However, as already mentioned above, the models can accurately describe migration in the past and that in the present respectively. Hence we see that the difference between the models illustrates an essential difference between migration in the past and that in the present. We investigate the essential differences between these models mathematically in the present research. In particular we are concerned with what causes these essential differences. In the following mathematical investigation we will employ advanced mathematical methods developed in Tabata and Eshima (2004).

Hotelling's model and the master-equation model do not consider population growth. However, in the real world, *relocation*, *growth*, *new entry*, and *exit* of population are a reality. It is natural to consider these kinds of behavior. In almost the same way as the models mentioned above, Hotelling makes a mathematical model of migration with population growth, which is described by the linear diffusion equation with a *source term* (see Hotelling (1978, p. 1234, (39-41)) and Beckmann and Puu (1990, p. 64, (9))). This equation is also known as the *Fisher-Kolmogoroff equation* in mathematical ecology (see Murray (2002, p. 400, (11.16-17))). Moreover we can construct a mathematical model with population growth, which is described by the master equation with the source term (see Tabata et al. (1999, p. 171, (E))). Hence, if we take population growth into account, then we need to compare those equations with the source terms. However, it is extremely difficult in mathematics to do this, because the necessary equations do not satisfy *the conservation law of population* that is essential to the methods developed in Tabata and Eshima (2004) (see Tabata and Eshima (2004, (3.18))). Hence, in order to avoid this mathematical difficulty, we consider only population relocation and we do not take growth, new entry, or exit into consideration. In the present research we will consider only Hotelling's linear diffusion equation model and the master-equation model (these models satisfy the conservation law of population).

In the last section we will prove the following theorem:

Theorem 1.1. (i) If the *cost of movement* required in migrating is high, then Hotelling's model and the master-equation model are very close to each other. (ii) If the cost of movement required in migrating is low, then the models are very distant from each other.

Recalling that Hotelling's model and the master-equation model can accurately describe migration in the past and that in the present respectively, and making use of the above theorem, we present the primary result of this research as follows:

Main Result 1.2. An essential difference between migration in the past and that in the present is caused by the difference between the cost of movement in the past and that in the present.

Furthermore, by Theorem 1.1, (i), we see that the master-equation model is applicable not only in the present but also in the past, because the cost of transit was high in the past. Moreover, it follows from Theorem 1.1, (ii), that Hotelling's model is not applicable in the present in general, because the cost of movement required in moving is low in the present by virtue of modern transit options. However, in some parts of the world, e.g., in Africa or Middle Asia, human mobility is still very costly. Hence, even in the present, we can apply Hotelling's model to migration in such situations.

In the next section we consider the master-equation model. Following this we consider Hotelling's model. In the same way as Hotelling (1978), we derive a nonlinear partial differential equation, which is called *the nonlinear diffusion equation*. The equation is called *the Fokker-Planck equation* in statistical mechanics (see, e.g., Risken (1996) and Okubo (1990)). Furthermore, we derive it also from the master equation. We obtain the linear diffusion equation through a linearization of the master equation. After this derivation we prove Theorem 1.1 in the final section.

2. The Master-Equation Model

As mentioned in the preceding section, the master equation can provide a description of migration in the present. The equation has the following form (see Helbing (1995, (1.5b), (4.1a))):

$$\partial f(t,x)/\partial t = -w(t,x)f(t,x) + \int_{y\in D} W(t;x|y)f(t,y)dy, \tag{2.1}$$

where $f = f(t,x)$ is an unknown function, which represents the population density at time $t \geq 0$ at a point $x \in D$. By D we denote a bounded domain included in the 2-dimensional Euclidean space. We assume that human population relocate in D. In the present research, for simplicity, we assume that D is a 1×1 square such that

$$D := [0,1]\times[0,1]. \tag{2.2}$$

The integration kernel $W = W(t;x|y)$ represents the transition probability of population from $y \in D$ to $x \in D$, and the coefficient $w = w(t,x)$ is defined as follows:

$$w = w(t,x) := \int_{y\in D} W(t;y|x)dy. \tag{2.3}$$

The first term of the right-hand side of (2.1) represents the sum of all migration flow out of the point $x \in D$ into all the points $y \in D$. The second term of the right-hand side of (2.1) represents the sum of all migration flow into the point $x \in D$ out of all the points $y \in D$.

It is deduced in Helbing (1995, Chapter 7) and Weidlich and Haag (1988, (4.1)) that the kernel has the following form:

$$W = W(t;x|y) := \mu(t)\exp\{\mathrm{U}(t,x) - \mathrm{U}(t,y) - \mathrm{V}(x,y)\}, \tag{2.4}$$

where $\mathrm{V} = \mathrm{V}(x,y)$ is a given function of $(x,y) \in D \times D$ which is called the *cost of movement*, $\mathrm{U} = \mathrm{U}(t,x)$ is a function which represents the *utility* at time $t \geq 0$ at a point $x \in D$, and $\mu = \mu(t)$ is a function of the time variable which represents the *spatial flexibility* of population. The cost of movement $\mathrm{V} = \mathrm{V}(x,y)$ is a sociodynamical quantity that represents a cost incurred in moving from $y \in D$ to $x \in D$, the utility denotes a sociodynamical quantity representing economic potential, and the flexibility is a sociodynamical quantity that represents an activity of population (see Helbing (1995) for the details of these sociodynamical quantities).

It is deduced in Weidlich and Haag (1988, pp. 82-83, (4.15-19)) that the utility depends on the following socioeconomic and economic variables (u_i), $i = 1,\ldots,5$: (u_1) *unemployment rate*, (u_2) *export structure index* (industrial minus agricultural export divided by the total export), (u_3) *overnight stays per capita*, (u_4) *percentage of total employment in the tertiary sector*, and (u_5) *population density*. In the same way as Tabata et al. (2000, 2001, 2002a, 2002b, 2004), we treat the variable (u_5) as an *endogenous* variable, and regard the other variables (u_1-u_4) as constants independent of the time and space variables. We assume that the utility depends only on the population density in the same way as Tabata et al. (2000, 2001, 2002a, 2002b, 2004), i.e., we make the following assumption:

Assumption 2.1. $\mathrm{U} = \mathrm{U}(t,x)$ has the following form: $\mathrm{U}(t,x) = U(f(t,x))$, where $U = U(\cdot)$ is a sufficiently smooth function.

It is deduced in Weidlich and Haag (1988, pp. 82-83, (4.14)) that the flexibility depends on the following socioeconomic and economic variables (v_i), $i = 1,\ldots,4$: (v_1) *total job vacancies*, (v_2) *total income*, (v_3) *investment structure index* (expansionary minus rationalizing investment divided by total investment), and (v_4) *total population*. As already mentioned in the preceding section, the total population (v_4) is constant. The variables (v_1-v_3) are regarded as *exogenous* variables in Weidlich and Haag (1988). In the present research, for simplicity, we assume that the variables (v_1-v_3) are constants independent of the time and space variables, as in Tabata et al. (2000, 2001, 2002a, 2002b, 2004). Hence, the flexibility is identically equal to a positive constant. The positive constant can be made equal to unity through a suitable choice of the time unit as in Beckmann and Puu (1990, p. 60). Hence we will assume that the flexibility $\mu = \mu(t)$ is identically equal to 1; there is no loss of generality.

It is natural to assume that if the distance between a point x and a point y becomes long, then the cost of movement required in moving from y to x becomes high, and that if the distance between x and y becomes short, then the cost of movement required in moving from y to x decreases. Hence, we assume that $V(x,y)$ is a monotonically increasing function of $|x-y|$. In order to investigate how the difference between the master-equation model and Hotelling's model changes according to the cost of movement, we will introduce a positive-valued parameter into the cost of movement calculation such that the cost of movement becomes higher as the parameter becomes smaller. Therefore we make the following assumption:

Assumption 2.2. $V(x,y) = |x-y|/\varepsilon$, for each $x,y \in D$, where ε denotes a positive-valued parameter.

From this assumption we see that if $\varepsilon > 0$ becomes small, then the cost of movement required in moving becomes high, which makes it difficult for a population to move further away. Moreover we deduce that if $\varepsilon > 0$ becomes large, then the cost of movement required in moving decreases, which facilitates distant moves.

By making use of Assumptions 2.1-2, we can rewrite the master equation (2.1) as follows:

$$\partial f_\varepsilon(t,x)/\partial t = -\, w_\varepsilon(t,x) f_\varepsilon(t,x) + \int_{y\in D} W_\varepsilon(t;x|y) f_\varepsilon(t,y)dy, \tag{2.5}$$

$$W_\varepsilon = W_\varepsilon(t;x|y) := \exp\{U(f_\varepsilon(t,x)) - U(f_\varepsilon(t,y)) - |x-y|/\varepsilon\}, \tag{2.6}$$

$$w_\varepsilon = w_\varepsilon(t,x) := \int_{y\in D} W_\varepsilon(t;y|x)dy, \tag{2.7}$$

where we attach the subscripts of ε to the integration kernel $W_\varepsilon = W_\varepsilon(t;x|y)$, the coefficient $w_\varepsilon = w_\varepsilon(t,x)$, and the unknown function $f_\varepsilon = f_\varepsilon(t,x)$ in order to emphasize that they depend on the parameter ε. We will consider *the initial-*

value problem for the master equation (2.5) with the following initial condition:

$$f_\varepsilon(0,x) = f_0(x),\ x \in D, \tag{2.8}$$

where $f_0 = f_0(x)$ is a positive-valued sufficiently smooth given function of $x \in D$. The initial-value problem has a unique solution, which is positive-valued and sufficiently smooth. No boundary condition is necessary for the master equation, since the unique solution of the master equation can be determined with no boundary condition (see Tabata et al. (1998, 1999, 2000, 2001, 2002a, 2002b, 2002c, 2004) for the mathematical details). The master-equation model is defined by the initial-value problem.

3. HOTELLING'S MODEL

We will construct Hotelling's model by following the line of Hotelling (1978, pp. 1229-1230) and Beckmann and Puu (1990, pp. 60-61). The population density is regarded as the driving force behind migration in Hotelling (1978) and Beckmann and Puu (1990). Hence we assume that the vector of migration has the following form, in the same way as Beckmann and Puu (1990, pp. 60-61):

$$G = G(t,x) := -\gamma \nabla g(t,x), \tag{3.1}$$

where we denote the population density at time $t \geq 0$ at a point $x \in D$ by $g = g(t,x)$. We write $G = G(t,x)$ as the vector of migration (at time $t \geq 0$ at a point $x \in D$) whose direction is that of the flow of migrants and whose length is equal to the rate of flow through a unit cross section per unit of time. The equation (3.1) contains negativity since migration is assumed to be in the direction of decreasing density. In order to avoid mathematical difficulties, Hotelling (1978) and Beckmann and Puu (1990) assume that γ is a positive constant (see Hotelling (1978, p. 1230, (19-20))). However, in the present research, we have no need to accept such an assumption. We have only to assume that γ is a positive-valued known function of the population density $g = g(t,x)$. Hence we write γ as follows:

$$\gamma = \gamma(g(t,x)), \tag{3.2}$$

where $\gamma = \gamma(\cdot)$ is a sufficiently smooth positive-valued given function. It follows from the conservation law of population that the thinning out of local population is equal to the thickening of the flow. Hence, in the same way as Beckmann and Puu (1990, pp. 60-61), we have

$$\partial g(t,x)/\partial t = -\nabla \cdot G(t,x). \tag{3.3}$$

By substituting (3.1) with (3.2) in (3.3), we obtain the following nonlinear partial differential equation (cf. Beckmann and Puu (1990, p. 61, (2-3))):

$$\partial g(t,x)/\partial t = \gamma(g(t,x))\Delta g(t,x) + \delta(g(t,x))\nabla g(t,x)\cdot\nabla g(t,x), \tag{3.4}$$

where we define

$$\delta(z) := d\gamma(z)/dz, \tag{3.5}$$

we denote the 2-dimensional *nabla operator* by ∇, i.e., $\nabla := (\partial/\partial x_1, \partial/\partial x_2)$, and Δ is the 2-dimensional *Laplacian*, i.e., $\Delta := \nabla\cdot\nabla = \partial^2/\partial x_1^{\,2} + \partial^2/\partial x_2^{\,2}$. Here we write the i-th component of x by x_i, $i = 1,2$. The equation (3.4) is the nonlinear diffusion equation (recall Section 1).

Next we will derive the nonlinear diffusion equation from the master equation (2.5). We can transform (2.5) into the following form as in Tabata and Eshima (2004, (2.9)):

$$\partial f_\varepsilon(t,x)/\partial t = F_\varepsilon(f_\varepsilon(t,\cdot))(x) + \mathrm{F}(\varepsilon,t,x), \tag{3.6}$$

where $\mathrm{F} = \mathrm{F}(\varepsilon,t,x)$ is a sufficiently smooth function of $(\varepsilon,t,x) \in (0,+\infty)\times[0,+\infty)\times D$ depending on the unknown function $f_\varepsilon = f_\varepsilon(t,x)$, and $F_\varepsilon = F_\varepsilon(\cdot)$ is a nonlinear partial differential operator of the following form (see Assumption 2.1 for $U = U(\cdot)$):

$$F_\varepsilon(k(\cdot))(x) := \alpha(\varepsilon)\{\beta_1(k(x))\Delta k(x) + \beta_2(k(x))\nabla k(x)\cdot\nabla k(x)\}, \tag{3.7}$$

$$\beta_1(z) := (1/4)\{1 - 2z\,dU(z)/dz\}, \tag{3.8}$$

$$\beta_2(z) := d\beta_1(z)/dz, \tag{3.9}$$

where $\alpha = \alpha(\varepsilon)$ is a positive-valued function of ε defined as follows:

$$\alpha(\varepsilon) := \varepsilon^4 \int_{0 \le x_i \le 1/\varepsilon,\ i=1,2} |x|^2 e^{-|x|} dx. \tag{3.10}$$

Dropping the residual term $F = F(\varepsilon,t,x)$ formally in the right-hand side of (3.6), we obtain the following nonlinear partial differential equation:

$$\partial g_\varepsilon(t,x)/\partial t = F_\varepsilon(g_\varepsilon(t,\cdot))(x), \tag{3.11}$$

where we denote an unknown function by a new symbol $g_\varepsilon = g_\varepsilon(t,x)$, because the equation thus obtained is different from (3.6). We easily see that if the following equality holds:

$$\gamma(z) = \alpha(\varepsilon)\beta_1(z), \tag{3.12}$$

then the equation (3.11) is the same as (3.4). We have thus derived the nonlinear diffusion equation from the master equation. In what follows we assume that (3.12) holds.

The equation (3.6) is equivalent to (2.5). If the residual term $F = F(\varepsilon,t,x)$ is sufficiently small, then the unknown function $g_\varepsilon = g_\varepsilon(t,x)$ is close to the unknown function $f_\varepsilon = f_\varepsilon(t,x)$. If not, then they are distant from each other. In the next section we will investigate whether $F = F(\varepsilon,t,x)$ is sufficiently small or not.

We need to assume that $\gamma = \gamma(z)$ is positive-valued in order that (3.4) is *uniformly parabolic* (see, e.g., Ladyzenskaja et al. (1968, p.11) for the uniform parabolicity), because if (3.4) is not uniformly parabolic, then we cannot deal with it at all. Hence, making use of (3.12) and (3.8), we assume that

$$1 - 2z\,dU(z)/dz \ge c_1, \tag{3.13}$$

where $c_1 \in (0,1)$ is a positive constant independent of $z > 0$. It follows from the differential inequality (3.13) that

$$U(z) \le \{(1 - c_1)/2\}\log z + U(1) \text{ for each } z \ge 1, \tag{3.14}$$

$$U(z) \ge \{(1 - c_1)/2\}\log z + U(1) \text{ for each } z \in (0,1]. \tag{3.15}$$

From this inequality we see that there exist a sufficiently large number of functions which satisfy (3.13). Hence, (3.13) is not restrictive.

We will seek a solution of the nonlinear diffusion equation (3.11). First we impose the following initial condition on (3.11):

$$g_\varepsilon(0,x) = f_0(x),\ x \in D, \tag{3.16}$$

where $f_0 = f_0(x)$ is the same as (2.8). In addition to the initial condition (3.16) we need to impose a boundary condition on the nonlinear diffusion equation. If the nonlinear diffusion equation has no boundary condition, then we cannot determine a unique solution to the equation. In the present research, for simplicity, we seek a solution of the nonlinear diffusion equation that coincides with the solution of the initial-value problem for the master equation (2.5) in the boundary of D, i.e., we impose the following boundary condition on the nonlinear diffusion equation (3.11):

$$g_\varepsilon(t,x) = f_\varepsilon(t,x),\ (t,x) \in [0,+\infty) \times \partial D, \tag{3.17}$$

where $f_\varepsilon = f_\varepsilon(t,x)$ is the solution to the initial-value problem for the master equation (2.5), and ∂D denotes the boundary of D. If we do not make (3.16-17), i.e., if we attempt to seek a solution that is different from the solution to the initial-value problem for the master equation at the initial time $t = 0$ or in the boundary ∂D, then the mathematical complexity is greatly increased. Hence, for mathematical simplicity, we impose the conditions (3.16-17). It is proved in Ladyzenskaja et al. (1968, Chapter V, Section 6) that the mixed problem for (3.11) with (3.16-17) has a unique solution.

Let us derive the linear diffusion equation by linearizing the nonlinear diffusion equation (recall Section 1). In the real world we frequently observe that the population density is not quite apart from a constant equilibrium, and that the population density does not change very steeply in space, i.e., that the spatial derivatives of the population density is small. Hence it is reasonable to make the following assumption:

Assumption 3.1. (i) The unknown function $g_\varepsilon = g_\varepsilon(t,x)$ is contained in the sufficiently small neighborhood of a constant equilibrium, i.e.,

$$\sup_{t \geq 0, x \in D} |g_\varepsilon(t,x) - c_2| << 1,$$

where c_2 is a positive constant which represents the equilibrium.

(ii) $|\nabla g_\varepsilon(t,x)|$ is sufficiently small. i.e.,

$\sup_{t\geq 0, x\in D}|\nabla g_\varepsilon(t,x)| << 1.$

By making use of Assumption 3.1, (i), we can replace $g_\varepsilon(t,x)$ contained in the coefficients β_j, $j = 1,2$, by the positive constant c_2 in (3.11). Moreover, by Assumptions 3.1, (ii), we can regard the second term,

$\alpha(\varepsilon)\beta_2(c_2)\nabla g_\varepsilon(t,x)\cdot\nabla g_\varepsilon(t,x),$

as negligibly small. Hence we obtain the following linear partial differential equation from (3.11):

$$\partial h_\varepsilon(t,x)/\partial t = \alpha(\varepsilon)\beta_1(c_2)\Delta h_\varepsilon(t,x), \tag{3.18}$$

where we denote an unknown function by a new symbol $h_\varepsilon = h_\varepsilon(t,x)$, because the equation thus obtained is different from (3.11). The linear partial differential equation (3.18) is the linear diffusion equation (see Hotelling (1978, (20)) and Beckmann and Puu (1990, p. 61, (3))).

Consider the mixed problem for the linear diffusion equation (3.18) with the same initial and boundary conditions as (3.16-17), i.e., with the following initial and boundary conditions:

$$h_\varepsilon(0,x) = f_0(x),\ x\in D, \tag{3.19}$$

$$h_\varepsilon(t,x) = f_\varepsilon(t,x),\ (t,x)\in[0,+\infty)\times\partial D. \tag{3.20}$$

This mixed problem defines Hotelling's model. It is proved in Ladyzenskaja et al. (1968, Chapter V, Section 6) that the mixed problem has a unique solution which is very close to that of the mixed problem for (3.11) with (3.16-17), i.e., that

$$\sup_{0\leq t\leq T, x\in D}|g_\varepsilon(t,x) - h_\varepsilon(t,x)| << 1, \tag{3.21}$$

where T is a positive constant independent of $\varepsilon > 0$. Hence, we will obtain estimates for $|f_\varepsilon(t,x) - g_\varepsilon(t,x)|$ in the next section.

4. Proof of the Main Result

Combining (3.21) and the following proposition, we obtain Theorem 1.1:

Proposition 4.1. (i) If the cost of movement required in moving is high, i.e., if the parameter $\varepsilon \in (0,1)$ is small, then the solution $f_\varepsilon = f_\varepsilon(t,x)$ of the initial-value problem for (2.5) with (2.8) is close to the solution $g_\varepsilon = g_\varepsilon(t,x)$ of the mixed problem for (3.11) with (3.16-17) in the following sense:

$$\sup_{x\in D, 0\leq t\leq T}|f_\varepsilon(t,x) - g_\varepsilon(t,x)| \leq c_3\varepsilon^6, \tag{4.1}$$

where T and c_3 are positive constants independent of ε.

(ii) If the cost of movement required in moving is low, i.e., if the parameter $\varepsilon > 0$ is large, then the difference between the solution $f_\varepsilon = f_\varepsilon(t,x)$ of the initial-value problem for (2.5) with (2.8) and the solution $g_\varepsilon = g_\varepsilon(t,x)$ of the mixed problem for (3.11) with (3.16-17) increases in the following sense:

$$\lim_{\varepsilon\to+\infty}\sup_{x\in D, 0\leq t\leq T}|f_\varepsilon(t,x) - g_\varepsilon(t,x)| = +\infty, \tag{4.2}$$

where T is a positive constant independent of ε.

Proof. We will prove only (i), because (ii) can be obtained in the same way as (i). The following inequality is proved in Tabata and Eshima (2004, Lemma 3.5):

$$\sup_{x\in D, 0\leq t\leq T}|\mathrm{F}(\varepsilon,t,x)| \leq c_4\varepsilon^6 \text{ for each } \varepsilon \in (0,1), \tag{4.3}$$

where T and c_4 are positive constants independent of ε. From this inequality we see that if the cost of movement required in moving is high, i.e., if $\varepsilon > 0$ is small, then the residual term $\mathrm{F} = \mathrm{F}(\varepsilon,t,x)$ of the master equation (3.6) is small (recall that (3.6) is equivalent to (2.5)). Subtracting the nonlinear diffusion equation (3.11) from (3.6), recalling that $f_\varepsilon = f_\varepsilon(t,x)$ coincides with $g_\varepsilon = g_\varepsilon(t,x)$ at the initial time $t = 0$ and in the boundary ∂D (see (2.8), (3.16), and (3.17)), and making use of (4.3), we can obtain (i) (see Tabata and Eshima (2004, 2009) for the mathematical details).

A result similar to the above proposition is obtained also in physics by use of an infinite series called *the Kramers-Moyal expansion* (see, e.g., Risken (1996)). However, the calculations done in physics are not acceptable mathematically, since it is not proved that the Kramers-Moyal expansion is convergent (see Tabata and Eshima (2004, Remark 4.1) for the details).

ACKNOWLEDGMENTS

This work was supported by Grant-in-aid for Scientific Research, Ministry of Sciences and Education of Japan.

REFERENCES

Åndersson, A. E. and Philipov, D., (1982), Economic models of migration, in *Regional Development Modeling: Theory and Practice*, (M. Albegov, A. E. Åndersson, and F. Snickars, ed.), North Holland, Amsterdam.

Beckmann, M. J. and Puu, T., (1990), *Spatial Structures*, Springer-Verlag.

Haag, G. and Weidlich, W., (1986), A dynamic migration theory and its evaluation of concrete systems, Regional Science and Urban Economics, Vol. 16.

Helbing, D., (1995), *Quantitative Sociodynamics*, Kluwer Academic Publishers, Dordrecht, the Netherlands.

Hotelling, H., (1978), A mathematical theory of migration, Environment and Planning, Vol. A10, pp. 1225-1239.

Ladyzenskaja, O. A., Solonikov, V. A., and Ural'ceva, N. N., (1968), *Linear and Quasilinear Equations of Parabolic Type*, AMS Translations of Mathematical Monographs, Vol. 23.

Murray, J. D., (2002), *Mathematical Biology, Vol. I: An Introduction, Third Edition*, Springer-Verlag, N.Y.

Okubo, A., (1990), *Diffusion and Ecological Problems: Mathematical Models.* Springer-Verlag, N.Y.

Richard, G., (1996), *Ailleurs, L'Herbe est Plus Verte, Histoire des Migrations dans le Monde*, Panoramiques-Corlet.

Risken, H., (1996), *The Fokker-Planck Equation*, Springer-Verlag, Berlin.

Stroock, D. W. and Varadhan, S. R. S., (1997), *Multidimensional Diffusion Processes*, Corrected Second Printing, Springer-Verlag, Berlin.

Tabata, M., Eshima, N., Takagi, I. and Hiroyama, T., (1998), The Cauchy problem for the system of equations describing migration motivated by regional economic potential, Applied Mathematics and Computation, Elsevier Science, Vol. 94, No. 1, pp. 45-64.

Tabata, M., Eshima, N., and Takagi, I., (1999), The nonlinear integro-partial differential equation describing the logistic growth of human population with migration, Applied Mathematics and Computation, Elsevier Science, Vol. 98, No. 2, pp. 169-183.

Tabata, M. and Eshima, N., (2000), The behavior of solutions to the Cauchy problem for the master equation, Applied Mathematics and Computation, Elsevier Science, Vol. 112, No. 1, pp. 79-98.

Tabata, M., Eshima, N. and Takagi, I., (2001), The master equation approach to self-organization in labor mobility, in *Evolutionary Controversies in Economics*, (Y. Aruka ed.), Springer-Verlag, Tokyo, pp. 201-212.

Tabata, M. and Eshima, N., (2002a), Blowing up solutions to the Cauchy problem for the master equation, Applied Mathematics and Computation, Elsevier Science, Vol.127, No. 2-3, pp. 181-193.

Tabata, M. and Eshima, N., (2002b), The Cauchy problem for the nonlinear integro-partial differential equation in quantitative sociodynamics, Applied Mathematics and Computation, Elsevier Science, Vol. 132, No. 2-3, pp. 537-552.

Tabata, M., Ide, A., Eshima, N., Takagi, I., and Takei, Y., (2002c), A functional-analytic and numerical-analytic approach to nonlinear economic models described by the master equation, in *Empirical Science of Financial Fluctuations*, (H. Takayasu ed.), Springer-Verlag, Tokyo, pp. 304-311.

Tabata, M. and Eshima, N., (2004), The Fokker-Planck equation and the master equation in the theory of migration. IMA Journal of Applied Mathematics, Oxford University Press, Vol. 69, Issue 3.

Tabata, M. and Eshima, N., (2009), The Kramers-Moyal expansion of the master equation that describes human migration in a bounded domain, Nonlinear Analysis Series B: Real World Applications, Elsevier Science, Vol. 10, pp. 639-664.

Weidlich, W., (2000), *Sociodynamics*, Harwood Academic Publishers.

Weidlich, W. and Haag, G., (1983), *Concepts and Models of a Quantitative Sociology*, Springer, Berlin.

Weidlich, W. and Haag, G. (Eds.), (1988), *Interregional Migration*, Springer, Berlin.

Zhang, W. B., (1989), Coexistence and separation of two residential groups - An interactional spatial dynamic approach, *Geographical Analysis*, Vol. 21, pp. 91-102.

In: Urbanization
Editor: Harold J. Kauffman
ISBN: 978-1-63117-063-8

Chapter 4

SUSTAINING URBANIZATION IN A DEVELOPING ECONOMY: A CASE STUDY OF NIGERIA

Oluwayomi Kayode Babatunde*
Department of Building, School of Design and Environment,
National University of Singapore, Singapore

ABSTRACT

This study models the relationship involving globalization, urbanization, and sustainable development (SD) where SD is defined further as a function of sustainable urbanization. It repositions culture as one of the pillars of SD, underscores SD's operational definition as "complementarity" as against "uncompromising-ness" of efforts to meet the present and the future needs, and advances a performance-based SD index. This study delineates between the two terms "developing economies" and "emerging economies" in favor of the former. Accordingly, it adopts Nigeria as a case study of a developing economy grappling with urbanization (largely due to rural-urban divide). Nigeria is the fastest urbanizing country in Africa based on its population. Conversely, 80% of Nigeria's 50% urban population lives in slums, attributable to Nigeria's housing deficit of over 16 million units, which is

* Corresponding author: Oluwayomi Kayode Babatunde. Department of Building, School of Design and Environment, National University of Singapore, 4 Architecture Drive, Singapore 117566. E-mail: yomi.babatunde@gmail.com.

increasing annually. Identifying housing as a menace to as well as an indicator of Nigeria's urbanization, this study summarizes the failure of prior efforts on housing in Nigeria into not being CLASS (culturally-acceptable, livable, affordable, sociable, and scalable) as well as not being SMART (specific, measurable, achievable, relevant, and time-bound). Premised on a SMART CLASS solution, it proposes an open-loop urbanization model based on decentralized developments in Nigeria as against the prevailing focus on the capital states. With the increasing volume of foreign direct investment (FDI) in Nigeria and concerns of the benefits thereof, this study also investigates the role of foreign players in sustaining urbanization in Nigeria. Specifically, it underscores China's significant role as a development agent in Nigeria, its leading role in ISO standards on quality management and environmental management, and its exemplary role as the fastest urbanizing country in the world.

Keywords: China, Culture, Developing Economies, Housing, Nigeria, Urbanization, Sustainable Development, and Sustainable Urbanization

1. Introduction

Sustainable development (SD) pervades all aspects of human endeavors. On the one hand, the principles underpinning SD is innate to human and as such predates the advent of SD as a technical concept. On the other hand, the surge in environmental degradation and the depleting impacts on diversity has brought SD to the forefront every development. Pre-industrial revolution (prior 1760), urbanization was relatively unknown; during industrial revolution (1760-1850), urbanization was necessary; post-industrial revolution, urbanization has been increasing unabatedly. Globalization, as typified by the industrial revolution, has fueled and continues to fuel urbanization, albeit now in the form of technological revolution. With finite resources, home operations have been expanded beyond national frontiers so much so that McLuhan (1964) declared that the world has become a global village. Perhaps, either a village without governance or a village with extended colonialism to the extent that considerations for meeting present needs lend no credence to the future needs. The erstwhile implicit concept of *social inequality* (attributable to Karl Marx (1818-1883) and Max Weber (1864-1920)) began to play out pervading, notably, cultural, social, economic, and environmental aspects of any development. The trend persisted until the Brundtland's (1987) report asserted that technology and social organization could both be managed and improved to allow for a new era of economic growth.

Famously, the report posited, "humanity has the ability to make development sustainable to ensure that it meets the needs of the present without compromising the ability of the future generations to meet their own needs". Following, it has become imperative to measure SD along the cultural, social, economic and environmental aspects. Conversely, due to different endogenous and exogenous factors, not all economies have the technology to ensure SD is implemented, successfully. And developed economies and their agents are leveraging on the inefficiencies of the developing economies to advance their own operations while leaving the latter with huge debt and depleted regenerative capacities. Asia, Africa and Latin America constitute the significant portion of developing economies with Africa facing the biggest challenge with regard to its development due to its being, characteristically, the poorest region (see Asiedu, 2002).

Nigeria's population is the highest in Africa (about 164 million people as of 2011 according to UNFPA, 2011). Nigeria is the fastest urbanizing country in Africa (about 50% according to Oluwakiyesi, 2011). Nigeria's urban tradition dates back centuries and is showing no sign of slackening in this century (see Rotberg, 2013). About 80% of Nigeria's urban dwellers were living in slums in 2007 (see Giddings, 2007) and, as of date, Nigeria's housing deficit has been estimated to be over 16 million units in deficit (see Ashkin, 2013). Past efforts on Nigeria's urbanization crisis have not been holistic and, in consequence, have been non-actionable. This study's aim is to investigate how to sustain urbanization in Nigeria. The three objectives include investigating: (1) the three key issues of public housing policy, which include the quantity of houses, the type of properties required to meet the needs of modern family types, and ensuring an acceptable standard of living; (2) the roles of major foreign players and how these foreign players can complement efforts towards sustainable urbanization without compromising the four aspects of SD; and (3) environmental management system among foreign and local firms in Nigeria, identifying the gap, and recommending solutions in line with the other two objectives.

1.1. Methodology

This study adopts review of the relevant literatures in order to fill theoretical gap in the existing literatures thereby contribute to knowledge and frame the research conceptually (see Alston and Bowles, 2003; Scandura and Williams, 2000).

Unlike other research designs, there are few scholarly works expounding on literature review as a domain. As such, some fields of studies, notably engineering and information systems, have not made as much theoretical and conceptual progress (see Levy and Ellis, 2006).

Succinctly, "an effective review creates a firm foundation for advancing knowledge. It facilitates theory development, closes area where plethora of research exists, and uncovers areas where research is needed" (Webster and Watson, 2002: xiii).

This study also adopts, specifically, a type 1 case study through intensive study of a single unit (in this instance, Nigeria) for the purpose of understanding a larger class of units (in this instance, developing economies) (see Gerring, 2004; Rowley, 2002). Commonly, a case study is applied to explain complex causal links in real life interventions, to describe the real life context that the intervention has occurred, to describe the intervention itself, and to explore those situations that the intervention being evaluated has no clear set of outcomes (see Tellis, 1997).

1.2. Globalization, Urbanization, and Sustainable Development: The Relationship

Urbanization is a two-way process of population concentration including "the multiplication of points of concentration" and "the increase in size of individual concentration" until an "urban maturity" is reached (Tisdale, 1942). Urbanization started in the 18th century while *urbanized societies* (a high proportion of the population living in cities) started in the 19th and 20th centuries and increasing afterwards (Davis, 1955). Five decades later, "urbanization has become a somewhat unfashionable topic among archaeologists for being bound up with questions of state formation" (see Osborne, 2005). Still, urbanization has remained the process by which a rural area becomes urbanized such that *urban area* can be conceptualized as including towns, cities and associated infrastructure (see Gaston, 2010).

Globalization nurtures urbanization and, as it turns out, constitutes challenges for urbanization, especially in developing economies (see Achankeng, 2003). Globalization diminishes geographic distance to establish and maintain cross-border economic, political, and social-cultural relations (Lubbers and Koorevaar, 2000). This is because globalization encompasses the causes, course, and consequences of transnational and transcultural integration of human and non-human activities (Al-Rodhan and Stoudmann, 2006).

Hence, it is possible to have primary globalization (technological innovation and its inadvertent negative consequences) and secondary globalization (new symbiotic relationship of concerned parties for good governance towards SD (see Lubbers and Koorevaar, 2000).

Earlier, Brundtland (1987, clause 27) asserted that technology and social organization could both be managed and improved to allow for a new era of economic growth. More famously, Brundtland (1987) asserted "humanity has the ability to make development sustainable to ensure that it meets the needs of the present without compromising the ability of the future generations to meet their own needs". The latter assertion has become famous as a definition of SD.

Cautiously, Daly (1990: 2) delineated between SD and sustainable growth (considered as a bad oxymoron) to submit that Brundtland's accomplishment lies on providing a political opening for the proper concept of SD to evolve. Still, SD has remained vague to such a great degree that Sharifi and Behnoud (2013: 115) has submitted that it is [still] evolving. IISD (2012) has traced SD, as a phenomenon, to Rachel Carson's 1962 *Silent Spring* while CASI (2012) has traced the international prominence of SD to the 1972 UN Conference on the Human Environment in Stockholm.

Nonetheless, CASI (2012) and IISD (2012) concurred that the 1987 Brundtland report brought together social, economic, cultural and environmental issues and popularized SD [and the issues as pillars or aspects of SD]. Conversely, Drexhage and Murphy's (2010) post-Brundtland study revealed that SD has been limited to an environmental issue, compounded by the prevalent view of a development as purely economic growth.

Oddly, developed economies are adopting this distorted framework to maintain their wealth while developing economies are, increasingly, following the same course.

Summing from the foregoing discussions, globalization affects urbanization and urbanization affects SD. The impact of urbanization on SD is a measure of how urbanization is being sustained or otherwise.

Consequently, Figure 1 presents the conceptual relationship involving globalization, urbanization, and SD.

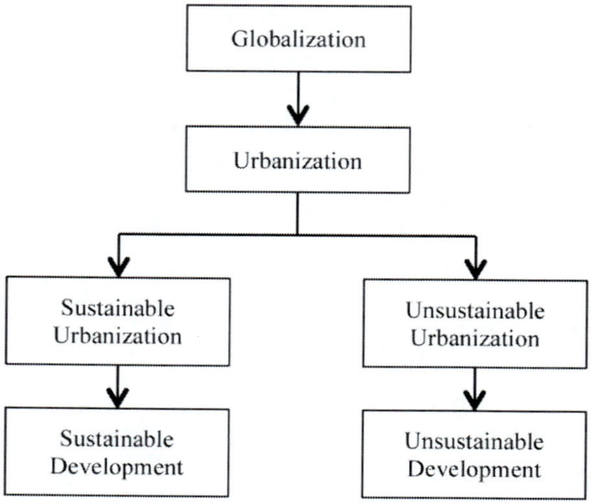

Figure 1. Conceptual Model: Globalization, Urbanization and Sustainable Development.

2. Sustainable Development as a Function of Sustainable Urbanization

The preparatory committee for the Rio+20 identified *green economy* in the context of SD and poverty eradication as well as *institutional framework for sustainable development (IFSD)* as the two main themes (see Leggett and Carter, 2012). The *green economy* upheld that each country would choose its own green economy approach and policy mix while the *IFSD* upheld that reform of existing international processes and institutions on the environment could benefit environmental protection programs. The *green economy* and the *IFSD* resonate with a performance-based approach to sustainability and an institutional-gap bridging respectively, both of which are supported by the findings from Drexhage and Murphy's (2010) study. As such, SD has been both an evolutionary concept as well as a sensitizing concept.

From Mebratu (1998), it can be deduced that while SD and sustainability rose to prominence simultaneously, sustainability, being prehistorically a key tenet of any human society, predates SD. Hall (2000a: 14) also attributed the advent of SD to the environmental degradation associated with development and failure of traditional approaches to solve the economic and social problems of the tropics.

From construction standpoint, Reckford (2006) noted sustainability as being a "green building" described as the process of building houses in a way that lessens the detrimental impact on the natural environment. Earlier, Harris (2003) submitted and expounded that the essential aspects of SD represent the elements of sustainability.

Thus, sustainability is an index for measuring SD so much so that Lubbers and Koorevaar (2000: 13) have identified the two levels of sustainability as the *weak sustainability* and *strong sustainability*. *Weak sustainability* involves conserving available resources [and] enabling future generations to achieve similar welfare levels while *strong sustainability* (more adventurous) involves conserving ecosystems in their present diversity. Measured on a weak sustainability, Pearce and Atkinson's (1993) study categorized Nigeria as "unsustainable".

2.1. Culture as One of the Discrete Pillars of SD

Despite being fundamental in the Brundtland's report (see IISD, 2012: 5), culture's role as one of the pillars of SD has, subsequently, been downplayed. Its explicit mentioning in SD literatures (for example, see Harris, 2003) has been perfunctory and this could be attributed to two reasons. The first is the dichotomous question of culture's existence and its commoditization (see Kopytoff, 1986) while the second is the simplistic notion of culture's subordination to technology (see Postman, 1992). Geert Hofstede (1928-), famous for the seminal work on culture, addresses the first reason in the sustained position that culture is a construct in the sense that it does not exist but rather made up [intangibly] to understand a complex world (see CIL, 2011).

In line with culture being a construct, Ferguson (1992: 69) has construed that culture, like globalization, is a myth in the sense that it is an ideological and explanatory approach to classifying sets of ideas. For the second reason, Hofstede et al. (2010: 391) have, likewise, flawed the notion as an illusion of the dominance of technology over culture. Similarly, Steve Jobs (1955-2011), famous for innovation in computer technology, has distilled technology as a tool for culture (see Isaacson, 2011: 84-6) in the sense of combination of the two as against the subordination of one over the other (see also Johnson, 2011; Lehrer, 2011). Jobs' position agrees with McSweeney (2009: 22) who, on the one hand, surmised to the causal (but not deterministic) influence of culture and, on the other hand, posited that all cultures are [arguably] incoherent.

Culture's incoherence stems from not having a single definition as noted by Sennara and Hartman (2002). Earlier work by Tylor (1871) defined culture as the complex whole, which includes knowledge, beliefs, art, morals, law, custom, and any other capabilities and habits acquired by man as a member of society. The all-inclusive definition by Tylor saw to Kroeber and Kluckhohn (1952) and Storti (1998) identifying 150 definitions and 300 definitions of culture respectively. As such, Hall's (1976) summation of culture as the way of life of a people pervaded the literatures. Consequently, Hofstede's (1980) study on cultural differences has shifted attention from anthropological definitions by defining culture as the collective programming of the mind, which distinguishes the members of one group from another. Still, authors have noted a common core in the different definitions (see Luthans and Doh, 2012: 108; Sennara and Hartman, 2002). Hence, Whiten and van Shaik (2007: 191) submitted that criteria exist to classifying a phenomenon as being cultural for detailed comparative analyses between two different cultures. According to Hofstede and Hofstede (2005: 23), a *dimension* is an aspect of a culture that can be measured relative to other cultures. And the four major frameworks for analyzing the many dimensions of culture include Kluckhohn and Strodtbeck's six dimensions, Hofstede's four [now six] dimensions, Trompenaars and Hampden-Turner's seven dimensions, and House et al.'s nine dimensions (see Ahlstrom and Bruton, 2010: 44-59).

Thusly, Karl Max's belief that technology determines economic structure that in turn determines social stratification and culture is misleading since economic systems are the superstructures; culture and history, the foundation; and politics and government, the core (see Wolgin, 1999: 16-7). Again, while technology increases access to information, it neither increases people's absorptive capacity nor change people's value systems since people select information based on their values and whatever reinforces the values (Hofstede et al., 2010: 391). In fact, culture has been found to be affecting technology adoption (see Lee et al., 2013) and significantly contributing to innovation (see Kaasa and Vadi, 2008). Following, the economic, social, and environmental pillars alone cannot reflect the complexity of current society since it is culture that shapes development and people's action (see UCLG, 2012). Culture enables effective development interventions; promotes sustainability inclusiveness, equity and diversity; and [which eventually] results in economic and non-monetized benefits (see UN System Task Team, 2012). As noted by Kavaliku (2005), the role of culture as a discrete pillar of SD transcends achieving immediate objectives to that of a quality of life befitting the integrity of everyone.

Daly (1990: 2-4) also underscored the importance of the complementarity of manmade and natural capital while Lubbers and Koorevaar (2000: 12-3) posited that SD has proved to be a sensitizing concept inviting people and organizations to think long term, inclusive and holistic. Hence, replacing "without compromising" in the Brundtland's 1987 definition with "and complements" can derive a more functional and operational definition of SD.

2.2. The Four Pillars of SD and SD Index

Advancing from the preceding section, the four pillars are complementary and of equal importance to achieving SD. This position supports Rio+20's two themes discussed earlier (see Section 2). Firstly, since culture, which shapes the economic and political systems of different countries, varies, it supports the *green economy*'s flexibility approach. Secondly, the equal importance being placed on the four pillars as against the prevalent focus on economic or environmental pillar, as Drexhage and Murphy (2010) have found out, supports *IFSD*'s institutional and process reforms. Consequently, pertaining to urbanization as conceptualized earlier (see Figure 1), measurements of the SD can be construed as an alpha sign as presented in Figure 2. The body or core represents the four pillars of SD and the intersection represents the SD index that determines the horn, which measures sustainable urbanization, and the tail, which measures unsustainable urbanization. Modeling it further in order to be able to quantify the SD index, each of the four pillars in the core can be depicted as a quadrant, which is equal to 90 degrees or 25 percent such that the sum of the core is equal to 360 degrees or 100 percent as presented in Figure 3.

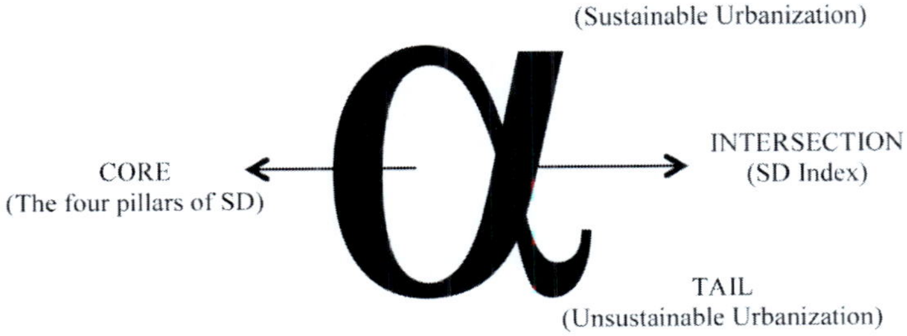

Figure 2. Sustainable Development as an Alpha Sign.

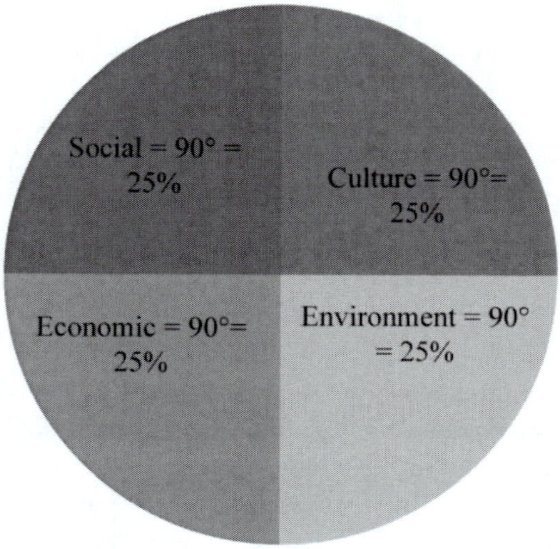

Figure 3. Quantifying the Four Pillars of Sustainable Development.

From Figure 3, the sum of the core serves as a measure of the SD index, which is equal to 360 degrees or 100 percent. Being an index, 100 percent can simply be expressed as 1 suggesting that the value of SD index ranges from 0 (unsustainable) to 1 (sustainable).

Furthermore, as presented in Figure 3, each pillar, being a quadrant, can be expressed as 90 degrees. However, degree measurements are arbitrary in trigonometric functions.

Thus, this necessitates radian measurements, which have the advantage of actual physical measurements (see Moore, 2013; Tuna, 2013 for discussions on degree and radian measurements). Formulas 1 and 2 present the conversions from degrees to radians and vice versa respectively.

$$1 \text{ degree} = 1 \times \pi/180 \text{ (in radian)} \quad (1)$$

$$1 \text{ radian} = 1 \times 180/\pi \text{ (in degree)} \quad (2)$$

Each quadrant represents the total of the number of variables identified under each quadrant. As such, irrespective of the number of variables that are identifiable under each pillar, the quadrant would still be equal to 90 degrees (or $\pi/2$ in radians). For example, each of the four frameworks for analyzing culture has different number of dimensions as discussed in the preceding sub-section.

Similarly, there are several frameworks to identify the variables under each of the other three pillars such as the ones provided by Gibberd (2008) and Lee (2012). It then follows that since each quadrant corresponds to 90 degrees, Formula 3 can be adopted to compute the *de facto* or equal value (D_V) of the different variables under the cultural, social, economic, and environmental pillars denoted as V_C, V_S, V_{Ec}, and V_{En} respectively.

$$D_V = (90 \text{ x } 1/N_T)\ \pi/180 \quad (3)$$

where, N_T is the total number of variables under each of the respective pillars.

In corollary, the value of each of the pillars (that is, V_C, V_S, V_{Ec}, and V_{En}) can be computed by multiplying the D_V under each of them by the number of the variables observed (N_O) using Formula 4.

$$V_P = D_V \text{ x } N_O \quad (4)$$

where, V_P represents each of the four pillars.

Hence, Formulas 3 and 4 can be combined to compute V_P for the respective four pillars by adopting Formula 5.

$$V_P = (90 \text{ x } N_O/N_T)\ \pi/180 \quad (5)$$

Premised on the foregoings, Table 1 presents the approach to the performance-based SD index developed in this study. A performance (and not a prescriptive) approach has been adopted since there are different frameworks to identifying the variables under each pillar. In addition, performance measurements allow for further enhancements and refinements of systems following their applications as supported by Neely et al.'s (2000) study. Importantly, the performance-based SD index is derived from the summation of the values of the four pillars as presented in Formula 6.

$$\text{SD index} = \sum [V_P] = V_C + V_S + V_{Ec} + V_{En} \quad (6)$$

From Table 1, the summation of the different values of V_P gives the performance-based SD index, which, in an ideal situation, equals to 2π radians (or 360 degrees). Recalling from Figure 3, 360 degrees is equivalent to 100 percent or, expressed simply as an integer, 1; thus revealing SD index as ranging from 0 (unsustainable) to 1 (sustainable) with a value less than 0 impossible, theoretically and practically.

Table 1. Performance-Based SD Index

No.	Pillars of SD	Variables of the pillars of SD (N_T)	De facto value of each variable (D_V)		Value of each pillar (V_P)
			Degrees	Radians	
1	Culture	$C_1, C_2, \dots C_n$	$(90 \times N_O/N_{TC})$	$(90 \times N_O/N_{TC})\ \pi/180$	V_C
2	Social	$S_1, S_2, \dots S_n$	$(90 \times N_O/N_{TS})$	$(90 \times N_O/N_{TS})\ \pi/180$	V_S
3	Economic	$Ec_1, Ec_2, \dots Ec_n$	$(90 \times N_O/N_{TEc})$	$(90 \times N_O/N_{TEc})\ \pi/180$	V_{Ec}
4	Environmental	$En_1, En_2, \dots En_n$	$(90 \times N_O/N_{TEn})$	$(90 \times N_O/N_{TEn})\ \pi/180$	V_{En}
SD index = $V_C + V_S + V_{Ec} + V_{En}$			X	Y	Z

Table 2. Computation of SD Index: Hypothetical Case

No.	Pillars of SD	Variables of the pillars of SD (N_T)	De facto value of each variable (D_V)		Value of each pillar (V_P)
			Degrees	Radians	
1	Culture	4	72	0.4π	0.20
2	Social	3	54	0.3π	0.15
3	Economic	2	36	0.2π	0.10
4	Environmental	1	18	0.1π	0.05
SD index			180	π	0.50

As an analogy, Table 2 presents a hypothetical case of a proposed development that meets four, three, two and one variable(s) of five possible variables under the cultural, social, economic and environmental pillars of SD respectively. Under the V_P, since each pillar equals to 25 percent (that is, 0.25), the D_V is, in consequence, expressed as fractions of the respective quadrants (for example, $0.4\pi / \pi/2 \times 0.25 = 0.2$).

Parties adopting the performance-based SD index can also set SD threshold (SD_T) to serve as an acceptable or agreed value for each of the pillars (between 0 to 0.25) or as a composite (between 0 and 1) for the four pillars below which a proposed development compromises than complements future needs. Following, values greater than or equal to the SD_T fall typifies the horn (in favor of sustainable urbanization) while values less than the SD_T typifies the tail (against sustainable urbanization). Interpretatively and as decision rules, the foregoing explanations with respect to the SD_T are presented in formulas 7 and 8 respectively.

$$\text{SD index} \geq SD_T = \text{Horn (sustainable urbanization)} \quad (7)$$

SD index < SD_T = Tail (unsustainable urbanization) (8)

In furtherance, the performance-based approach has the advantages of flexibility (framework for selection of variables), simplicity (easy to understand and undertake), and interpretability (easy to explain and translate into action). This study opines that basic understanding of SD should precede more advanced approaches to modeling and quantifying SD (for example, see Eriksson, 2013; Quaddus and Siddique, 2013).

On the other hand, some other existing qualitative tools for assessing SD (for example, see Gibberd's 2002 *sustainability building assessment tool* and Lee's 2012 *total building performance approach*) should advance towards quantification. The Management Consultant Peter Drucker (1909-2005) has succinctly and strategically construed "If you can't measure it, you can't manage it."

3. Developing Economies or Emerging Economies?

Unlike *developing economies*, the definition of *emerging economies* is elusive so much so that the World Bank maintains that it does not have a list of emerging markets or economies (see World Bank, 2013). Antoine van Agtmael coined the phrase *emerging markets* in 1981 when IFC was pitching an idea of a "Third World Equity Fund" to a group of leading investment managers. Agtmael (2007: 5) recounted that the term *third world* suggests stagnation while the term *emerging markets* suggests progress, uplift, and dynamism.

Two decades after, scholars have crystallized *emerging economies* to meaning low-income, rapid-growth countries using economic liberalization as their primary engine of growth (see Hoskisson et al., 2000). In addition, the term has been sub-divided into the two main groups of *developing economies* (in Asia, Latin America, Africa and the Middle East) and *transition economies* (the former Soviet Union and China). In the decade following, the need for a more practice-oriented understanding of the term *emerging economies* arose. Remarkably, Khanna and Palepu (2010) tackled the challenge by construing *emerging economies* as typifying transactional arenas where buyers and sellers are not easily or efficiently able to come together. Khanna and Palepu (2010) argued that "developing economies" fall short in a number of ways due to their institutional voids, which, thus, defines economies as emerging [from inefficiencies].

From the foregoing accounts, semantically, while a *developing economy* can be said to be emerging (that is, Khanna and Palepu, 2010), an *emerging economy* cannot be said to be developing (that is, Hoskisson et al., 2000). Acceptably, as crystallized by Hazy (1990), an economy can grow without developing, or develop without growing, or do both or neither.

Thus, this study disagrees with Radelet's (2010) study on seventeen emerging African countries to, operationally, restricting the term emerging economies to Brazil, Russia, India, China and South Africa (BRICS). Cooper (2006: 258) has defined BRIC (pre-South Africa) as large-population countries with relatively dynamic economies that could, if appropriate policies were pursued, occupy an increasingly important place in the global economy from 2003 to 2050.

Similarly, the Economist (2013) has succinctly explained that BRIC (formed in 2001) was in 2010 extended to include an African country that South Africa as against Nigeria would [better] keep the acronym BRICS intact. By and large, the term *emerging economy* is not elitist nor is the term *developing economy* derogatory. As surmised by Srinivasan (2009: 64), emerging market is a bit of both a creation and a discovery as well as a bit of both a description and a norm. Gaillard (2010) provides a more holistic categorization of the 115 developing economies encompassing the newly industrialized economies (neither listed as developed nor least developed).

3.1. Developing Economies: Perceptions, Realities and Opportunities

According to Magnusson and Wydick (2002), African markets are, often, neglected by international investors for being hard to penetrate, located in war-torn regions, [and] or located in sanctioned countries. Thus, African countries like the other developing economies have to adopt branding techniques. Radelet (2010: 28) rightly noted despite Africa's economic decline having ended since 1995, the continent's popular perception is that of a region of stagnation, conflict, and authoritarian rule.

Plausibly as a result, Radelet (2010: 28) branded some developing economies in Africa as emerging economies, which agrees with Agtmael's (2007: 5) submission that emerging markets was conceived as a "branding maneuver". Rotberg's (2013) study sheds light on Africa's current situations. Rotberg (2013: 5-6) posited that Africa in 2013 is much better led and much better governed than it was during the first post-colonial period.

However, Africa needs to embrace robust policies to overcome its deficiencies and harness all its capabilities and possibilities to launch into its next era of growth and development. On the one hand, being able to achieve the desired growth and development is premised on responsible leadership (see Rotberg, 2013: 189-215). On the other hand, leadership's degree of adherence to sustainable principles would determine the effectiveness and implementability of policies needed to enhance any anticipated development in Africa (see Odebiyi, 2010: 28).

Developing economies would account for about 55% of global construction by 2020 with China and India driving growth in emerging markets (GCP and OE, 2011: 8). Nigeria would register rapid growth from its rising population, housing, and infrastructure needs while South Africa would see growth in its booming infrastructure from growing private investment (GCP and OE, 2011: 10).

In addition, cities in the developing economies, particularly in Africa and Asia, would absorb most of the world population growth in the next 50 years (see IHC, 2009). Paradoxically, a relatively low standard of living, an undeveloped industrial base, and a moderate-to-low human development index score and per capita income characterize developing economies (Gaillard, 2010: 80). In many cities and towns in Africa, less than 10% of the people live in formal housing sector because urban population growth outpaces formal housing production (IHC, 2009: 11). This is being compounded by poor quality housing construction, overcrowding, and the lack of secure tenure for most informal sector housing (IHC, 2009: 11). Recalling, about half of Nigeria's population lives in urban areas making it one of the fastest urbanizing countries in Africa (Oluwakiyesi, 2011: 8). As of 2007, more than three-quarter of Nigeria's urban dwellers live in slums (see Giddings, 2007: 2), which can be attributed to the country's huge and mounting housing deficit (see Oluwakiyesi, 2011: 22).

3.2. Developments, Urbanization, Challenges and Opportunities in Nigeria

Abegunrin (2006: 235) viewed globalization as that resulting into the free movement of capital, goods, and services across national boundaries to identify *economic globalization* (economic activities), *global economy* (cross-national system), *global production* (resources), and *global finance* (global exchanges).

Playing out, Nigeria's oil boom has been traced to the first oil shipment in 1958 and the subsequent oil export dominance till the peak in 1980 after which oil prices began to weaken (Obidegwu, 1996: 126-7). Nonetheless, Nigeria's public sector has grown substantially during the oil boom from state's direct investment in industrial activities and state's enlarged bureaucracies to such a great degree that urbanization rose from 21.4% in 1972 to 27% in 1980 (see Obidegwu, 1996: 128-130). The oil boom's adverse effects thus necessitated the design of the structural adjustment program (SAP) (between 1986 to 1990 according to Oyejide, 1991), which, nonetheless, failed. SAP's failure was due to the then military government tenure's failure to demonstrate SAP's effectiveness to Nigeria's economic recovery and the subsequent face-off between the government and the labor unions (see Fashoyin, 1996). By 1991, the housing problem in Nigeria was 8 million units in deficit based on the finding of the New National Housing Policy (NNHP'91) initiated by the then military government to meet Nigeria's housing problems by 2000 (see Madaki and Ogunrayewa, 1999). Succeeding SAP, the post-military administration's homegrown strategy of National Economic Empowerment and Development Strategy (NEEDS) has reformed macroeconomics, structural systems, public sector, as well as institution and governance (see Okonjo-Iweala and Osafo-Kwaako, 2007). NEEDS has progressed in its effort to address structural and institutional weaknesses of Nigeria's economy as well as tackle corruption and overhaul public expenditure management (AfDB and OECD, 2008: 510). As a result, Nigeria's real GDP growth averaged 6.5% during the period 2003 to 2007 (AfDB and OECD, 2008: 510) and crossed the growth rate of 7% in 2011 with [an even more] innate potential to record higher growth (Oluwakiyesi, 2011: i). Nigeria's real estate sector is divided into the residential and non-residential sectors with the former accounting for the larger part of the country's real estate or housing demand (Oluwakiyesi, 2011: 22), where there is an enormous deficit. With NEEDS, privatization has increased in the housing sector through available finance from banks' capitalization. In parallel to Nigeria's GDP growth, the construction sector contributed about 1.3% of the GDP in 2006 (AfDB and OECD, 2008: 510) and rose to about 1.4% in 2010 (Oluwakiyesi, 2011: 2). Driving the growth is Nigeria's sustained urban tradition, which Rotberg (2013: 27-8) noted, unlike the other African countries', dates back centuries and since increasing rapidly and endlessly. As it turns out, the hard choice for developing economies is to temper consumption, resource use, export more, and use domestic resources at home to foster long-term sustainability (see Brown et al., 2000: 713 for detailed discussions).

Africa is, and would continue to be for a long time to come, an agrarian society; thus, underscoring that questions relating to land and access to it are [and would continue to be] crucial (Konadu-Agyemang and Panford, 2006: 6). Nigeria and South Africa are the leading economic powers in Africa (see Abegunrin, 2006: 246; The Economist, 2013). Nonetheless, it would be difficult for these two African countries or any African country to achieve SD without significant conscious internal efforts complemented by those of the developed economies. As such, Abegunrin's (2006) study concluded that if Nigeria's myriad of struggles continue, the world and particularly Africa would be worse off in the 21st century. Similarly, Radelet's (2010) study concluded that if South Africa's economy takes a turn for the worse, all bets are off [Africa as a viable investment destination]. James (1999: 3) has posited rightly that when efficiency is the operative concept of developments, economies are able to maximize the use of their resources and minimize waste and destruction. As a result, the strategic decision taken in shaping, managing and using urbanization as well as its products (towns, cities and megacities) determines if it supports SD or otherwise.

4. Sustaining Urbanization in Nigeria

Existing literatures on housing, urbanization, and SD in Nigeria are fragmented. This can be attributed to inadequate effort to synthesize existing findings to identify the gap.

Hence, there is a need to align the disparate views and advance a SMART (specific, measurable, achievable, relevant, and time-bound, according to Doran, 1981) solution to address Nigeria's mounting urbanization crisis. Nigeria's population was about 164 million people in 2011 (see UNFPA, 2011: 119). The breakdown of the population for the 36 states and the federal capital territory (FCT), Abuja is presented in Appendix I. As revealed in Appendix I, save for the six new states (created post-1991), 25 states of the remaining 31 states have witnessed growths in urbanization above 40% between 1991 and 2011. Nigeria's total urbanization of about 50% suggests that about 82 million Nigerians were living in urban areas as of 2011. Incommensurably, as of mid-2013, Nigeria's housing deficit has been estimated to be about 17 million units and increasing annually by 2 million units (see Egene, 2013). As such, not only was the NNHP'91's target of "housing for all Nigerians" by 2000 missed (see Madaki and Ogunrayewa, 1999), Nigeria's housing deficit worsened by a 100% in two decades.

And if the growth rate in housing deficit continues as Egene (2013) has estimated, Nigeria will not achieve its millennium development goal (MDG) on housing as Orji (2013) has predicted.

As of 2012, there were 10.7 million houses in Nigeria and of which 9.6 million (about 90%) were "self-built", defined as constructing a permanent home from scratch by the owner rather than paying a professional builder to do it (Alagbe, 2013: 7, 14).

Conversely, about 80% of Singapore's population lives in public housing and about 90% own their flats (see HDB, 2013). While Singapore has about 3% of Nigeria's population (see UNFPA, 2011: 119), government policies, nonetheless, drive private housing prices (see Phang and Wong, 1997).

In addition, affordability of private housing in Singapore is limited by land scarcity so much so that the country has developed a strong housing program that provides quality low-cost housing (see Ong, 2000).

Operationally, UN-HABITAT (2003: 12) has defined a slum as an area deficient in any of the five characteristics including *safe water*, *sanitation and other infrastructure*, *structural quality of housing*, *overcrowding*, and *security of residential status*. As such, Alagbe's (2013) definition fell short of distilling that some private housing in Nigeria have also been contracted to unqualified/ incompetent and/or unethical professionals.

This has, in the least, contributed to dysfunctional buildings and, at worst, ultimate building collapse with a significant effect on SD (see Windapo and Rotimi, 2012). By and large, government's lack of provision of the necessary infrastructure mars private housing alike and degenerates them into slums.

4.1. An Institutional Perspective on Housing Developments in Nigeria

According to Lowe (2011: 2), the three key issues of housing policy include: (1) the quantity of houses (current and forecast), (2) the type and size of properties that meets the needs of modern family, and (3) an acceptable standard of living. In addition to failing housing policies (see Aribigbola, 2013), Nigeria still grapples with literacy, notably, in financial sustainability (see Wolgin, 1999: 11-6 for detailed discussions).

For example, following Nigeria's USD 18 billion (about 60%) Paris Club debt relief in June 2005 (see Okonjo-Iweala, 2007), its external debt has again risen to USD 5.7 billion in 2011 (see Top Asia News, 2012).

As Brown et al. (2000: 713) have rightly flagged, developing economies' unsustainable consumption leads to debt, loss of productive capacity of their environments, and inability to pay back their debt (amongst competing needs). Hence, while Lowe (2013: 18) posits that social housing fills the gap between housing supply and demand, Nigeria's housing policy lacks social housing provision (see Okparaocha, 2013). Social housing is related to economic restructuring (Lowe, 2013: 230), which is just consolidating in Nigeria. Nigeria's past social housing schemes have failed (see Aribigbola, 2013; Dodo, 2004) weakening government's ability to meet the housing deficit to the thriving of self-build housing. In consequence, Kabir and Bustani (2009) have submitted that Nigeria's housing problem relates to both quantitative and qualitative inadequacies.

In response to the World Bank's January 2013 decry of the "poor quality of statistics and data in Nigeria" (see Abasiekong, 2013), the current housing deficit in Nigeria was about being ascertained as of July 2013 (see Premium Times, 2013).

Similarly, the different effort to combat Nigeria's housing deficit has seen to the institution of the [embroiled] National Housing Fund (NHF) Act that supports employers' 2.5% deduction of employees' basic salaries (see Ogunwusi, 2013); the Federal Housing Authority's (FHA) acquisition of about 4,100 hectares of land spanning nineteen states (Kolawole, 2013); the World Bank's support to provide 200,000 low-cost housing units in five years (see Udo, 2013); and the newly instituted World Bank soft credit-assisted Mortgage Refinancing Company's provision of cheaper mortgage loans from the year 2014 (see Egene, 2013).

Nigeria's Federal Ministry of Lands, Housing and Urban Development has also formed partnership with a consortium of Nigerians overseas to construct 10,000 housing units in each of the six geo-political zones (Pepple, 2012: 22). Similarly, Nigeria's government has also canvassed support from private investors as exemplified by the USD 6 billion housing sector investments from two Chinese firms (see Aid Data, 2013). A separate memorandum of agreement worth USD 1 billion was also signed with another Chinese firm in 2009 for the delivery of mass housing in the FCT, Abuja (see Aid Data, 2009). Chinese investments in Nigeria have been increasing steadily since 2006 as presented in Figure 4. Nigeria is China's second largest export market in Africa [after Angola as supported by CCS, 2009a; 2009b] (Osagie, 2011). Chinese investments in Nigeria covers housing, which poses unique challenges to foreign investment (see Brock, 2013) demonstrating China's competiveness in Nigeria.

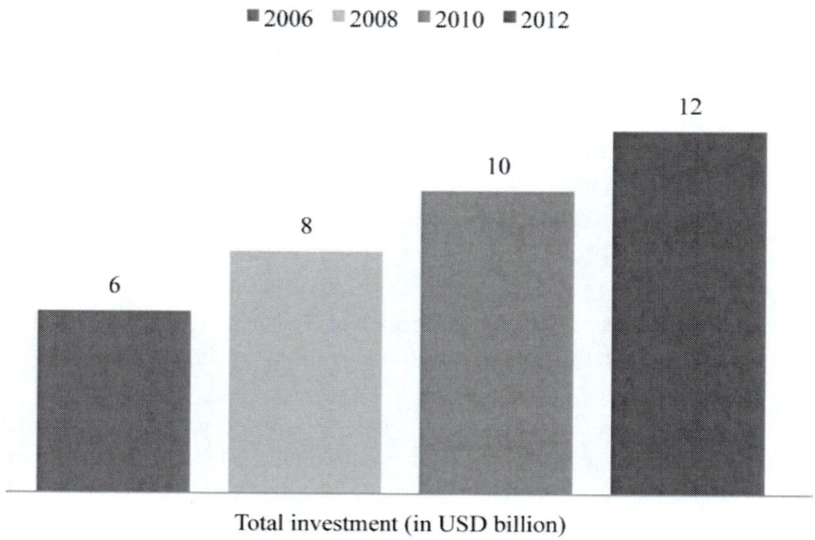

Figure 4. Chinese total investment in Nigeria between 2006 and 2012.

This is underpinned by China's track record in the delivery of public housing as typified by its five-year plan to complete 36 million public housing units by 2016 (see CCS, 2012). Strategically, another Chinese firm has partnered with Ogun state in Nigeria to deliver 2,000 units of public housing (PM News, 2013).

Still, there is a significant lag in Nigeria's housing supply (from huge deficit in housing units) and housing demand (from affordability of the housing units) (Alagbe, 2013: 7; Oluwakiyesi, 2011: 22). As of date, the conservative cost of a home in Nigeria is about N3.5 million [USD 22,055] (Alagbe, 2013: 7), average or modest cost is N5 million [USD 31,506] (Ashkin, 2013) while a decent home would cost a minimum of N10 million [USD 63,012] (Orji, 2013). On such a striking realization, Okparaocha (2013) stressed proper understanding of the terms *sustainable housing*, *social housing*, and *low-cost housing* in Nigeria.

Operationally, *social housing* can be (and often is) *low-cost housing* and both might not be *sustainable housing*. *Social housing* is related to economic restructuring, *low-cost housing* is not defined only in the initial cost (hence, the all-inclusive term "affordable housing") and *sustainable housing* is assessed both in *social housing* and *low-cost housing*. In sum, sustainable housing is a technical construct or term, while social housing and low-cost housing are social-economic constructs or terms.

4.2. A New Lens for Housing Development in Nigeria

Quality housing (residential and non-residential) is fundamental to urbanization. When there is a significant lag in housing, infrastructures break down and slums with related social problems thrive making urbanization unsustainable. Comparatively, while Nigeria aims to deliver 200,000 housing units in five years (see Udo, 2013), Singapore aims to deliver 204,461 housing units in three years (see Woo, 2013) and China aims to deliver 36 million housing units in five years (see CCS, 2012). Singapore's initiative is a joint effort including the government and the private sectors while China's initiative is underpinned by its track record in public housing delivery to cater to the needs of its over one billion population people. As discussed in the preceding sub-section, past housing programs in Nigeria have failed (Dodo, 2004) due, summarily, to quantitative and qualitative inadequacies (Kabir and Bustani, 2009). From the UK experience, as expounded by Lowe (2013: 22), the huge imbalance in the supply and demand of housing stems from the following, which are explained subsequently with respect to the experience in Nigeria:

a Demographic factors such as the growth of single-person households, divorce rates and remarriages;
b Unforeseen circumstances such as natural disasters, ethnic and militant violence and economic crises;
c The prevalent housing units such as existing housing units (size, location and ease of conversion to meet new needs; and
d The prevalent housing policies such as government's provision of housing, government's intervention through land use planning system and subsidies for housing repairs and renovation.

Firstly, from CIA (2013), Nigeria has the 8th largest population in the world and growing at the rate of 2.54%. Youth (15-24 years) and working (25-64 years) population constitute about 19.3% and 53.2% respectively. Sex ratio of the population is 1.01 male per female and total fertility rate is 5.31 children per woman.

Nigeria has over 250 ethnic groups and speaks over 500 indigenous languages in addition to English (official) and the three major languages (Hausa, Yoruba, and Igbo).

Secondly, in 2012, Nigeria's worst flooding in decades affected about 7 million people (see IRIN, 2013) with 31 states projected to experience at least a devastating flood in 2013 (see Akinsuyi and Sanni, 2013).

Similarly, communal clashes and militant attacks are increasing and impacting on the number of internally displaced people who are being forced into temporary housing where their needs are neither assessed nor addressed by the government (see IDMC, 2013).

Thirdly, a significant portion of the housing in Nigeria is self-built housing (recall Alagbe's 2013 definition) and as a result deficient in a way, which qualifies most of the self-built housing as a slum based on the UN-HABITAT's 2013 operational definition. Corroborating the foregoing position, Kabir and Bustani's (2009) study found that about 87% of the existing housing stocks in Nigeria do not meet the minimum quality requirements. (Recall, about 90% is self-build housing.)

Fourthly, this is a multi-faceted concern that, albeit receiving attention from the Nigerian researchers, has, nonetheless, defied significant progress. And there is a shared responsibility in this instance because, as highlighted earlier, existing literatures are fragmented.

Thus, policy makers are wallowing in a plethora of recommendations. Harris and Chib (2011: 175-6) have rightly construed that there is a limited relationship between academic impact and socioeconomic impact.

On the reverse, policy makers are also lacking the impetus to take action, which can be attributed to bureaucratic inertia of the government, rooted in culture (see Section 2.1). Table 3 summarizes the different recommendations from some of the notable studies on housing in Nigeria.

4.3. Developing a Model for Sustaining Urbanization in Nigeria

As can be deduced from Table 3, the different views have merits that if synthesized can give a holistic view of how to sustain Nigeria's rising urbanization.

Often, authors view SD as a fuzzy concept (see Rau and Fahy, 2013: 4-5) and, as a result, lean towards addressing the peripheral issues.

A fuzzy problem is characterized by imprecision and uncertainty thereby requiring an algorithm, which, adaptively, responds to the dynamism of the problem (see Konar, 2000; Pratihar, 2008). SD pervades all aspects of private and public activities that it has, seemingly, become a cliché.

A complete view of SD is defined in terms of the four pillars, which can also be quantified using, for example, the performance-based SD index. Globalization precedes urbanization, which must be guided by SD to derive a sustainable urbanization (see Figure 1).

Table 3. Past recommendations for housing developments in Nigeria (selective)

No	Author(s)	Recommendation
1	Madaki and Ogunrayewa (1999)	Having a holistic system on housing development that covers implementation, monitoring, and evaluation.
2	Dikko (2002)	Developing cost models by organizing historical information into accessible databases for realistic housing programs.
3	Olayiwola et al. (2005)	Encouraging private home building through local technology and provision of serviced pieces of land to individuals.
4	Ojo et al. (2006)	Varying procurement methods with project values since the former impacts the latter in terms of time and cost slippages.
5	Akeju (2007)	Moving away from the anomaly of government providing housing to addressing challenges against private and foreign investments.
6	Coker et al. (2007)	Gearing public- and private-directed improvements towards new towns and enforcing minimum housing standards.
7	Osasona (2007)	Designing public housing activities to suit ethnic needs rather than approaching housing needs as being culturally homogenous.
8	Nyam (2008)	Providing the necessary infrastructure (by the government) and partners' understanding of their roles and expectations in a private finance initiative (PFI).
9	Odebiyi (2010)	Adopting a hybrid of traditional and modern architectural practices for sustainable housing development.
10	Adeogun and Taiwo (2011)	Endorsing community participation in implementation, execution and allocation of the affordable housing units in a public-private partnership (PPP).
11	Ibem et al. (2011)	Creating an enabling environment by addressing contextual and organizational challenges in public housing provision.
12	Jiboye (2011a)	Instituting responsive housing policies in congruence with the existing national and socio-economic realities in Nigeria.
13	Jiboye (2011b)	Ensuring concerted effort towards an operational framework that supports sustainability of existing and new housing programs.
14	Windapo and Rotimi, (2012)	Revamping policies to set out minimum sustainability performance requirements for building and construction works.
15	Alagbe (2013)	Tailoring policies towards the needs of the poor who are in the majority to make housing affordable to more Nigerians.
16	Alufohai (2013)	Placing priority on restructuring the mortgage sector, identified as a major impediment to the goal of housing for all.
17	Aribigbola (2013)	Encouraging state and the local governments' participation and making housing policy inclusive and realistic.

Nigeria's soaring urbanization is not only being challenged by a huge deficit in housing and related infrastructure, it is equally being challenged by lack of maintenance of the existing units (see Odediran et al., 2012).

Different factors have been identified as being responsible for the maintenance debacle (see Waziri and Vanduhe, 2013). In sum, past and prevalent approaches to housing provisions in Nigeria have not been holistic and; thus, have led to inhumane and apathetic housing, which are unsupportive of "livable cities" both at the unit level (see Aduwo et al., 2012) as well as the neighborhood level (see Ukoha and Beamish, 1996).

Chen and Thwaites (2013: 1) have construed that culturally and socially responsive urban design practices would be more sustainable than the prevailing approaches, which dwell on architectural statements of power and prestige. This applies in Nigeria where Take (1984: 54) has long noted that it is difficult to find "Nigerian architecture" due to the destruction of much of the traditional buildings.

Hence, Konadu-Agyemang and Panford (2006: 6) have flagged that Africa's resource management problems are festering due to the abandoning of indigenous resource management systems in favor of Western ones. Conversely, indigenous knowledge is neither a primitive nor a naive knowledge reflective of an inferior stage of human cultural progress completely irrelevant now (Appiah-Opoku, 2006: 205).

Following, while Osasona (2007) stressed that Nigeria's indigenous traditional architecture should be true to itself to be sustainable, Odebiyi (2010) upheld that it could still be a hybrid system and be sustainable. The hybrid system is complementary and has been identified as a viable option to address Nigeria's housing deficit (see Agbedo, 2013; Uroko, 2013).

Complementarity concords with Tessema et al.'s (2009: 11) view of a sustainable building and construction as fulfilling its performance requirements with minimum adverse environmental impacts, while encouraging improvements in economic, social and cultural conditions at local, regional and global levels.

The building sector is one of the most important consumption clusters in terms of environmental burden with activities accounting for about 40% of overall energy use and associated greenhouse gas emissions (Tessema et al., 2009: 8). A reason is that while economic development drives resource utilization, it is not being linked with environmental integrity especially in resource-rich developing economies (James, 1992: 2). Urban development grapples with *material availability*, *city planning*, *applicable regulation*, *land availability*, *suitable staff*, and *financial challenges* (see Queiros, 2008).

And rural development grapples with the opportunity cost of permitting developments to conserving renewable and non-renewable resources (see James, 1999: 3). As such, confronting Africa's endowments is its production targets against rapid population growth and unpredictable climatic conditions (see also Konadu-Agyemang and Panford, 2006: 4). The foregoing dichotomy calls for a balancing act as exemplified by Hall's (2000b: 45-6) distilling of SD as "development" for the proponents and "sustainable" for the opponents. As a result, Hall (2000c: 728) rationalized that the only strategy to achieve sustainability is to align expectations with the capabilities of the resources to deliver them. In corollary, economic development models must be developed that tackle the negative social impacts caused by ill industrialization and physical infrastructural developments.

In developing such a model for Nigeria and which can be extended to other developing economies alike, the following four points must be borne on mind:

a Firstly, it has to be acknowledged that the country's prevailing model of urbanization is unsustainable as typified by its huge and rising housing deficit coupled with inadequate infrastructures due also, summarily, to the country's systemic and structural inefficiencies (see Aniekwu, 1995).

b Secondly, for each subsequent approach to urbanization to be sustainable, it must be all-inclusive by being "Culturally-acceptable", "Livable", "Affordable", "Sociable", and "Scalable" (herewith henceforth, CLASS). A synthesis of the underlying themes of Nigeria's urbanization problems, as have been discussed in the preceding sections, derives the mnemonic CLASS. *Cultural acceptability* for ethnic diversity; *Livability* for housing satisfaction (neighborhood level); *Affordability* for life-cycle cost; *Sociability* for housing satisfaction (unit level); and *Scalability* for planning and policy.

c Thirdly, it follows that, qualitatively, SD is characterized by being CLASS, which also supports the notion of a smart city. According to Giffinger et al. (2007), a smart city is well performing in a forward-looking way in the six characteristics, which include *economy*, *people*, *governance*, *mobility*, *environment* and *living*. Similar to Gibberd's (2002) sustainable building assessment tool (SBAT) and Lee's (2012) total building performance (TPB) approach, CLASS supports the implementation of SD but differs from SBAT and TPB in that it works complementarily with the SD index.

d Fourthly, with CLASS in mind, a model for SD in Nigeria should treat urbanization (measured, technically, by urbanized areas comprising towns, cities and associated infrastructures according to Gaston 2010), as a biological cell that self-renews as presented in Figure 5.

As an emphasis, construing urbanization as a biological cell makes it easier to make theoretical and practical sense of the growth of towns and cities. This questions Hillier's (2009) position that "the complexity of cities seems to defy description" with much left to be debated on how Hillier's concept of *space syntax* differs from a biological cell, save for ideological construct. In self-renewal, cells grow by taking on similar or different characteristics from the adult cells and the adult cells are responsible for maintaining the number of differentiated cells at a constant level (see Hall and Watt, 1989; He et al., 2009 for detailed discussions).

From Figure 5, Nigeria's 50% urban population can be self-renewed through one of two alternative scenarios of either focusing on decongesting the urban areas (scenario 1) or concurrent development of the urban and the rural areas (scenario 2).

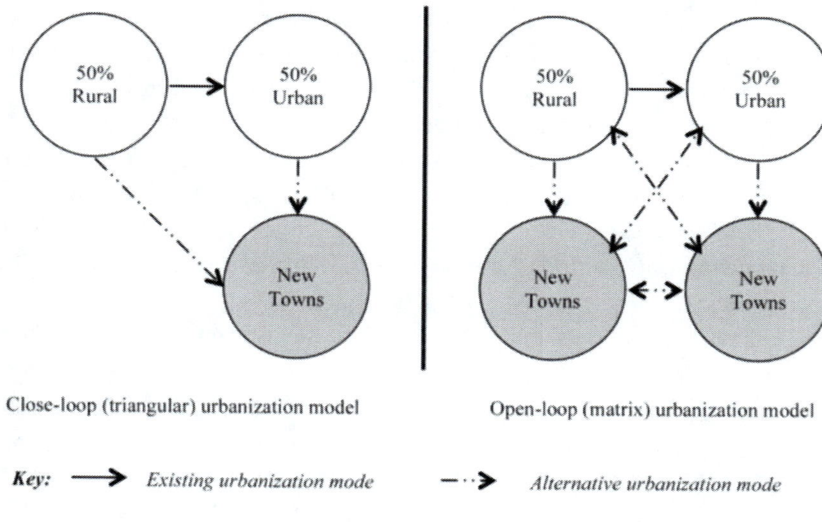

Figure 5. Sustaining Urbanization in Nigeria.

Scenario 1 typifies a close-loop or triangular urbanization model while scenario 2 typifies an open-loop or matrix model. The close-loop urbanization model resonates with Tisdale's (1942) process of reaching urban saturation point followed by stabilization or subsidence since it is inherent in any society to urbanize itself continuously. The open-loop urbanization model resonates with Davis' (1955) process of deriving a new kind of urban existence by spreading "rurality" evenly to be urbanized and industrialized. Hence, while the open-loop model is able to sustain itself, the close-loop system simply shifts stresses from one urban area to another neglecting rural areas.

The close-loop model promotes rural-urban divide (see Mason, 1989) leading to what Wolgin (1999: 9) has termed "urban hell", resulting from [a massive] migration to cities [often] without jobs thereby promoting the ills of urbanization.

Premised on the sustainable open-loop model, which can be operationalized by concurrently developing all the 36 states and the FCT (assuming 37 states in total), Table 4 presents specific, measurable, achievable, relevant, and time-bound (SMART) assessments of Nigeria's housing deficit. From findings from the preceding sections, Table 4 adopts the estimate of 17 million housing units in deficit, the target to deliver 200,000 housing units in five years, and the conservative, modest, and decent housing cost scenarios of USD 22,055, USD 31,506 and 63,012 respectively.

Table 4. SMART Assessments of Nigeria's Housing Deficit (17 million units)

No	Description	Public	Public and Private	Public, Private and Others
1	Housing deficit (in million)	17	17	17
2	Housing target (units in 5 years)	200,000	400,000	500,000
3	Housing delivery rate (units per annum)	40,000	80,000	100,000
4	Housing delivery per state per annum	1,082	2,163	2,703
5	Meet housing deficit (in years)	85	42.5	34
6	*Cost to meet housing deficit (in USD)	Conservative	Modest	Decent
		375 billion	536 billion	1 trillion

*Each cost is applicable to the sole-, joint- and multi-party arrangement(s).

As revealed in Table 4, if housing delivery effort was left solely to the government, it would take Nigeria about 85 years to meet its current housing deficit (and this is without factoring in growth in population and internal displacements, notably, from conflicts and flooding). This could be halved in a joint-party effort involving the private (foreign investors or local investors) and reduced further through a multi-party arrangement involving the Nigerian government, local investors and foreign investors. Taking housing deficit to be constant, the advantages of increased target and delivery rate as well as reduced time to meet the deficit makes a point for joint-party and multi-party arrangements such as the public-private partnerships in Nigeria (see Adeogun and Taiwo, 2011; Nyam, 2008). More so, comparatively, the moderate cost of meeting Nigeria's housing deficit is nearly the value of all the oil Nigeria has pumped since its independence [in 1960] (see Ashkin, 2013). And the decent cost of meeting the housing deficit equivalent to China's financing to Africa in the forms of investments and loans in the twelve years starting from 2013 (see CCS, 2013a). Contrastingly, it would take China less than 5 years to meet Nigeria's current housing deficit based on China's current housing delivery rate (see Section 4.2). As such, Nigeria is partnering with China on major housing projects as highlighted earlier (see Section 4.1). Unsustainably, the projects have focused on Abuja (the FCT) to the neglect of the other states; thus, gradually, shifting migration trajectory towards the FCT to a large extent that Abuja has experienced the highest population growth in twenty years (please refer to Appendix I) and fast-approaching its maximum capacity of 3 million people (see Take, 1984). Concentrating developments in Abuja or, notably, Lagos state (see Oyebade, 2013) is analogous to the unsustainable triangular or close-loop model. Alternatively, under the matrix or open-loop model as presented in Table 4, equal importance is given to all the states for concurrent developments. For example, the housing delivery per state per annum would translate into developments at the local government areas (LGAs) of the 37 states as opposed to just the state capitals. (Please refer to Appendix I for details of the LGAs and state capitals.)

4.4. Applying the Model: A Case Study of Lagos and Abuja in Nigeria

Technically, Nigeria operates two capitals, which includes a commercial capital and an administrative capital, the former being Lagos and the latter Abuja.

Lagos state was Nigeria's federal capital until December 1991 that Abuja became the FCT. According to Nigeria's National Bureau of Statistics, Lagos state, created in May 1967 by virtue of State Decree No. 14, is the smallest in area, the most populous [after Kano state; as can be gathered from Appendix I] and the most economically important of Nigeria's states. Lagos state became an administrative entity in April 1968 with Lagos serving as both the State and the Federal capitals.

Subsequently, the pro-July 1975 government inaugurated committee's study concluded that a new capital was to be created and built in a secure place, ethnically neutral, centrally accessible, and [with] sufficient land resources for urban development (Take, 1984: 49). Following, the government designated a FCT in the center of the country and established the Federal Capital Development Authority (FCDA) to be in charge of building a new capital city to be called Abuja (Take, 1984: 49). Figure 6 presents Nigeria's map, which shows Abuja at the center and Lagos at the southwestern part.

Figure 6. Map of Nigeria.

Following, FCDA drew up Abuja's development guidelines to accommodate a population of 1.6 million people by 2000 and a maximum population of 3 million people after which further population growth would be accommodated in satellite towns. Critically, Abuja was construed to be solely administrative with no effort to diversify its economic base. On the reverse, Abuja has experienced an astronomical growth in population, quintupling itself in 20 years to reaching 2.19 million people in 2011 (please refer to Appendix I). As Jibril and Garba (2012) have rightly noted, since becoming the new capital, Abuja and its environs have experienced unprecedented growth as a result of rapid influx of people from all over Nigeria and the world. Earlier, Ukoha and Beamish (1996: 27) have noted that the public housing conditions in Abuja were becoming poor quality from inadequate livable, affordable, and satisfactory housing. Ukoha and Beamish's (1996) study concluded that Abuja's mass housing scheme has under-represented junior-level government [and the unanticipated non-government] workers and their needs and, as such, missed critical attributes contributing to housing satisfaction. Similarly, Jibril and Garba's (2012) study concluded, albeit crediting the good intention, attributing the failure of the FCDA's scheme to the rush with which it has been implemented, the lack of cooperation among the relevant institutions, and Developers' violation of guidelines.

From the foregoings, Abuja is on its way to a rehash of Lagos experience. The shift of Nigeria's capital from Lagos to Abuja and the mounting urbanization crisis in Abuja is symptomatic of a close-loop urbanization model that Lagos has erstwhile embraced. Should Abuja become congested and blighted with the ills of urbanization, the possibility of another relocation cannot be ruled out and the opportunity cost to the other pertinent issues confronting Nigeria is unquantifiable. Large-scale urban development is inevitable in modern times (Chen and Thwaites, 2013: 1); thus, necessitating complementary actions that addresses present and future concerns. This is the premise of SD operationalized through the open-loop urbanization model characterized by CLASS. A developing economy must spread its urbanization evenly to close its rural-urban divide and conserve its natural resources to guard its regenerative capacities. According to the World Bank, Nigeria's rural population declined from 53.3% in 2000 to 46.7% in 2010 due to lack of basic infrastructures in the rural areas, which are still agrarian (see *Business Day*, 2013). Notably, Lagos, in its renewed capacity to sustain its development, has embarked on the provision of basic infrastructures in its rural areas (see *Business Day*, 2013).

Nonetheless, Nigeria should look beyond Lagos and Abuja drawing wisdom from the Greek philosopher Aristotle's (384-322 BCE) "the whole is greater than the sum of its parts."

5. The Role of Foreign Players in Sustaining Urbanization in Nigeria

Since economic development is the main catalyst for the utilization of natural resources, national economic policies should determine the rate and level of extraction of natural resources (see James, 1999: 2). In fact, economic development's environmental degradation with its associated multifaceted problems has brought SD to the fore (see Hall, 2000a: 14). The economic development of any economy is tied to the volume of its construction activities and can be expressed bi-directionally and cyclically (see Babatunde and Low, 2013a).

Consequently, scholars have concluded on the significant contribution of the construction industry to the economic development of different developing economies.

These include, *inter alia*, Indonesia (see Wibowo, 2009), Botswana (see Ssegawa, 1999) and Nigeria (see Oladinrin et al., 2012). Similarly, in attempts to bridge the gap in the development of the construction industries between the developing and developed economies, past studies have also investigated the state-of-the-art of the construction industry in developing countries (see Moavezandeh, 1976), the unique challenges of the construction industries in different developing countries (see Ofori, 2000) and a framework for revaluing of the construction industries in the developing countries (see Ofori, 2006). Africa is, and would continue to be for a long time to come, an agrarian society, which stresses that issues relating to economic development and SD would continue to be crucial.

Okpoko's (1998) archaeological study on Nigeria's urbanization concluded that ecology, centralized or quasi-centralized authority, and inter-regional trade have interacted in different ways to promote Nigeria's early urbanization.

Interestingly, Akeju (2007) attributed the failure of Nigeria's housing policy on affordable housing to "the unsustainable tenet that houses would be provided [solely] by government" to argue in favor of leveraging on involving the private sector as well as foreign investment.

Noting the huge potential in Nigeria's housing sector, Akeju (2007) expounded on challenges to be tackled to encourage foreign direct investment (FDI) to aid in the provision of affordable housing in Nigeria (see also Brock, 2013). Filaba (2008: 11-2), on the flip side, opposed FDI in Nigeria because government's coercive creation of enabling environments for foreign players is not being matched with significant developments from the foreign players. Earlier, Isimbabi (1999: 34) had also objected to opening up local markets due to investors' exploitation. Filaba (2008) argued against "urban-urban" and "rural-urban" migrations (characteristic of the close-loop urbanization model) noting they facilitate capitalism. Hence, Abegunrin (2006: 235-6)'s hard-to-fault position that globalization increases profit through operations in dirt-wage, cheap labor, low or no regulation underdeveloped countries instead of high-wage, high-regulation developed countries.

Foreign players also lack indigenous knowledge, which Konadu-Agyemang and Panford (2006) defined as localized knowledge unique to a particular society or ethnic group. Developing economies have also been disillusioned that indigenous knowledge is primitive and irrelevant to modern challenges as noted by Appiah-Opoku (2006). This supports Filaba's (2008) conclusion on Nigeria summarizing the strategy as make-believe of pre-colonial insignificance in order to establish a new system for exploitation. Similarly, it echoes Drexhage and Murphy's (2010) position of developed economies' distorted framework to maintain their wealth with developing economies tailing the course.

Brown et al. (2000) have noted this unsustainable system leaves developing economies with hard-to-repay debt due to the loss of productive capacities of their environments. Still, the results of FDI vary across economies (developed, developing and transition); spurring and crippling economic growth in some and having spill over benefits in some (see Görg and Greenaway, 2003; Moran et al., 2006).

Among the developing economies, in Africa, neither infrastructure development nor openness to trade has a significant impact on the flow of FDI (see Asiedu, 2002). Conversely, FDI in infrastructure responds positively to effective domestic regulatory frameworks (see Kirkpatrick et al., 2006) as well as subscribing to international trade agreements (see Büthe and Milner, 2008) to increasing economic growth.

Thusly, there is a shared responsibility between host economies and foreign investors. And so, James (1999: 3) opined efficiency should be the operative concept of developments in Africa to attain equilibrium between development and conservation.

Furthermore, in many African countries, the ability to prioritize development objectives to sustain economic development is non-existent due to political instability and selfish gains (James, 1999: 3).

In this direction, culture plays an important role in an economy's development and implementation of objectives as supported by Harrison and Huntington's (2000) study that also contrasted Asian and African economies. A decade earlier, Hofstede and Bond's (1988) study had contrasted Asian and Western economies and concluded likewise that culture in the form of certain dominant values is a necessary condition for economic growth. IBM Institute for Business Value's (2013) study has also identified culture as an important driver in value creation because it influences advocacy and action on the information available.

Similarly, with the right culture, private and government entities can embrace the responsibility of emphasizing the sustainability of natural resources, while encouraging and fostering capacity building, which empowers citizens with the ability to practice environmentally sensitive development (see James, 1999). In this regard, global governance alike has a significant role to play as discussed subsequently.

5.1. Global Governance

Global institutions (amongst other, the UN, IMF, World Bank and WTO) are responsible for global governance. Finkelstein (1995) defined global governance as governing relationships, without sovereign authority, beyond national frontiers. Subsequently incoherent use of the term global governance saw to Dingwerth and Pattberg (2006) submission that it is "a heuristic device to capture and describe the confusing and seemingly ever-accelerating transformation of the international system". Murphy and Yates (2009: xii) noted the International Organization for Standardization (ISO) as a global institution being a key player assisting to advance and shape patterns of industrialization.

In addition, Murphy and Yates (2009: xiii) saw ISO as a non-governmental institution bridging public and private sectors and as the self-proclaimed international standard setter for business, government and society, through its pursuit of voluntary standards.

A *standard*, according to Crawford (2007: 208) is a measure, devised by general consent as a basis for comparison, against which judgments might be made as to levels of acceptability.

To have effect, standards do not need to be officially endorsed as they can be voluntarily accepted (see Crawford, 2007: 208; Murphy and Yates, 2009: 8-11). As noted by Murphy and Yates (2009: xiii), ISO has developed more than 17,000 international standards since being founded in 1946 and continuously developing more standards yearly.

Organizations are certified to standards they subscribe to and are, subsequently, audited for compliance to ensure quality assurance and, by extension, continuous improvement.

Hence, Murphy and Yates (2009: 7) stressed that standards shape competition, which leads to innovation.

Recalling, culture impacts technology adoption (Lee et al., 2013) and innovation (Kaasa and Vadi, 2008) and now, as it relates to the ISO, culture impacts on union membership (see Posthuma, 2009). ISO's niche within global governance is its view of itself as a leader in building the "technical infrastructure" that is needed for SD and for achieving the UN MDGs (see Murphy and Yates, 2009: 47-50). And of ISO's numerous standards, the standards on quality management system (QMS), environmental management system (EMS) and occupational health and safety (OHSAS) have received the most acceptance, worldwide. Respectively, the QMS, EMS and OHSAS are referred to as ISO 9001, ISO 14001 and ISO 18001.

Collectively, in the forms of different combinations of the three systems, they are referred to as integrated management systems (IMS) or management system standards (MSS) (see Hortensius, 2013).

Increasingly, ISO 26000 (Social Responsibility) is becoming more popular as it responds to a general problem of industrial capitalism (see Murphy and Yates, 2009: 82-88). ISO 26000 provides guidance rather than requirements; thus, meaning, unlike the other three standards, it cannot (as of date) be certified to. Nonetheless, it clarifies what social responsibility is and how organizations can achieve this globally (see ISO, 2010).

5.2. ISO's Standard on Environmental Management System

Following ISO's publication of the quality management standard series (ISO 9000) in 1987, ISO built on its experience by creating the environmental management series (ISO 14000) in 1996 (see Murphy and Yates, 2009: 68, 77). Since 1993, ISO has been publishing a yearly survey of ISO 9000 certification, which saw to the inclusion of the ISO 14000 certification from 1999 (see ISO, 2012).

ISO 14001 is reputed as the world's most recognized framework for EMS that helps organizations to manage the impact of their activities on the environment and to demonstrate sound environmental management (ISO, 2009: 6). As of end 2012, 1,883,138 ISO 14001 certificates had been issued across 167 countries spanning the world's seven continents (see ISO, 2012). Table 5 presents the growth in the number of ISO 14001 certificates across the seven continents since 1999.

By industrial sector (see ISO, 2012), the construction sector has constituted the highest proportion of the ISO 14001 certificates issued between the years 2009 to 2012. This typifies a paradigm shift from construction as a mere economic activity to construction as a sustainable activity, encompassing all the four pillars of SD. Tessema et al. (2009: 11) have construed sustainable construction as meaning responsible supply, operation and maintenance of buildings that meet the needs of their owners and users over their lifespan with minimal unfavorable environmental impacts, whilst encouraging economic, social and cultural progress.

Table 5. ISO 14001 Certification by Continents

Year	Africa	Central/ South America	North America	Europe	East Asia and Pacific	Central and South Asia	Middle East	Total
1999	129	309	975	7253	5120	114	94	13994
2000	228	556	1676	10971	8993	267	156	22847
2001	311	681	2700	17941	14218	419	194	36464
2002	418	1418	4053	23305	19307	636	303	49440
2003	626	1691	5233	30918	25151	927	450	64996
2004	817	2955	6743	39805	38050	1322	862	90554
2005	1130	3411	7119	47837	48800	1829	1037	111163
2006	1079	4355	7673	55919	55428	2201	1556	128211
2007	1096	4260	7267	65097	72350	2926	1576	154572
2008	1518	4413	7194	78118	91156	3770	2405	188574
2009	1531	3748	7316	89237	113850	4517	2775	222974
2010	1675	6999	6302	103126	126551	4380	2515	251548
2011	1740	7105	7450	101177	137335	4725	2425	261957
2012	2109	8202	8573	113356	145724	4946	2934	285844
Total	14407	50103	80274	784060	902033	32979	19282	1883138

Adapted from ISO (2012).

In February 2013, ISO launched an online survey to understand the needs of users and other interested parties of ISO 14001 in a bid to revise ISO 14001:2004 for an updated version poised to be effective from the year 2015 (see Frost, 2013).

As of June 2012, more than 250,000 organizations around the world were using ISO 14001:2004 to improve their environmental performance (see Marlow, 2012). Overall, as noted by IEMA (2013), the number of ISO 14001 certifications jumped 9% in 2012 (see also ISO, 2012 for the executive summary of the ISO 2012 Survey), following ISO's confirmation that the world's leading EMS would be revised. From Marlow (2012), following includes the views on the future of ISO 14001:

- a The current version of 14001 is a flexible and an effective standard; nonetheless, additions and enhancements are needed;
- b Organizations would link EMS to operational strategies so that environmental considerations are accounted for earlier and across organizations;
- c Organizations would consider the impacts of changing environment on their businesses and vice versa for business resilience [and sustainability];
- d Organizations would be allowed to evaluate progress through customized key performance indicators that demonstrate continual improvement;
- e Maturity matrices would be introduced as a guidance tool to help organizations evaluate and plan the development of their EMS; and
- f EMS is [and should continue to be] a key aspect of the role of organizations [in a similar way that ISO 9000 QMS has matured].

Orts and Spigonardo's (2012) study opined that sustainability challenges and opportunities call for greening the supply chains, which is, similarly, the premise of ISO 14001 in its addressing not only the environmental aspects of an organization's processes, but also those of its products and services (ISO, 2009: 6).

ISO (2004) has also stressed that ISO 14001:2004 (revised of the first edition ISO 14001:1996) is applicable to any organization that wishes to: (1) establish, implement, maintain and improve EMS, (2) assure itself of conformity with its stated environmental policy, and (3) demonstrate conformity with ISO 14001:2004 by:

a Making a self-determination and self-declaration, or
b Seeking confirmation of its conformance by parties having an interest in the organization, such as customers, or
c Seeking confirmation of its self-declaration by a party external to the organization, or
d Seeking certification/registration of its EMS by an external organization.

As supported by Hortensius' (2013) study, ISO yearly surveys reveal that small and medium scale enterprises (SMEs) are increasingly implementing MSS (management system standards) based on the ISO 9001 and 14001. And the main motivation, according to the same study, is that many SMEs are suppliers to large companies that impose quality and environmental requirements on their supply chain.

Earlier, Delmas' (2001) study has also established that the involvement of external stakeholders in a firm's ISO 14001 standards creates competitive advantage, which is difficult to imitate by competitors. Expounding on the benefits derivable, Orts and Spigonardo (2012: 2-5) underscored the following as the current best practices to maximizing the roles of suppliers in "greening the supply chain", effectively:

a Choosing the right suppliers;
b Putting effort where it would do the most good;
c Collaborating to innovate;
d Using suppliers as force multipliers; and
e Ensuring compliance

5.3. Foreign Trade Partners at the Frontier of Sustainable Development in Nigeria?

From Isimbabi (1999: 29), developing economies' perpetual dependence on aid would not solve their long-term economic development due to donor countries' increasing aid-fatigue and anti-foreign aid sentiment. For this reason, developing economies should gravitate to attracting capital from private sources (domestic and foreign), premised on the institution of a well-functioning banking and financial system (see Isimbabi, 1999: 29-30).

As such, Nigeria's banking reforms (see Okonjo-Iweala and Osafo-Kwaako, 2007) have improved confidence in the country in recent times.

From Corporate Guides (2010: 46), Nigeria receives the largest amount of FDI in Africa including from traditional oil companies from the US (Chevron, Texaco and Exxon Mobil) and the UK (Shell).

More recently, oil and non-oil companies from China have joined the league with other significant sources of FDI including Italy, Brazil, the Netherlands, France, and South Africa. China is the largest development aid provider to developing economies (see UNDP, 2011: 147) and has Nigeria as its second largest export market in Africa, after Angola (see Section 4.1).

Table 6 presents the top ten international contractors in Africa for 2013 based on contracting revenues generated in the year 2012 according to the *Engineering News-Record* (ENR) (see Reina and Tulacs, 2013: 4 for detailed discussions). The ten companies have operations in Nigeria with sixty percent of the companies originating from China (see Table 6). This is reflective of the increasing number of Chinese companies in Nigeria, which has tipped the erstwhile official 30 often reported (see AfDB et al., 2011: 13; Chinese Embassy, 2004) to becoming 200 in 2013 (see Aderinokun and Eze, 2013).

As of July 2013, the total of Chinese investments in Nigeria has reached USD 13 billion (see Aderinokun and Eze, 2013), changing the trend in Chinese total investments in Nigeria drastically by a 30%-increase between 2012 and 2013 (see Figure 4 presented earlier for comparison).

Table 6. ENR's Top Ten International Contractors in Africa for 2013

Rank		Company	Origin
2013	2012		
1	1	SAIPEM	Italy
2	2	China Communications Construction Group	China
3	4	Sinohydro Group Ltd.	China
4	3	VINCI	France
5	6	China Railway Group Ltd.	China
6	10	CITIC Construction Co. Ltd.	China
7	9	China State Construction Engineering Corp.	China
8	8	Bouygues	France
9	NA	Construtora Norberto Odebrecht	Brazil
10	NA	China National Machinery Industry Corp.	China

Adapted from Reina and Tulacs (2013: 4).

Recalling the impact of MMS based on ISO 14001 and 9001 certificates on the supply chain spanning different SMEs (preceding sub-Section), ISO 14001 and 9001 certificates issued and withdrawn among Nigeria's top trading partners in the ten years prior to 2013 are presented in Table 7. As can be affirmed from Table 7, Nigeria had the least number of both certifications, which is reflective of the generally low adoption of quality certification among the country's SMEs (see Osagie, 2012).

More critically, Nigeria is, historically, small-scale industry oriented (see Ayozie, 2011), spanning even the construction industry (see Adams, 1997).

Nevertheless, while Nigeria had the lowest number of certificates, comparatively, it can be deduced that Nigeria had the least ratios of certificates issued to certificates withdrawn. This suggests the capability to sustain certifications among organizations in Nigeria once the inertia of being certified has been overcome. (Parenthetically, see Babatunde and Low, 2013b for issues relating to ISO 9001 certification in Nigeria and recommendations to adopting the certification as a platform to an all-encompassing QMS.)

Overall, Table 7 reveals a generally low adoption of EMS in comparison to QMS. Firstly, this can be attributed to QMS being premised on designing and delivering service that satisfies established "customers' needs"; whereas, EMS is premised on fulfilling legislated and regulated "environment's needs" (see BSI, 2013).

Table 7. ISO 9001 and 14001 Certifications among Nigeria's Top Trading Partners

No.	Country	ISO 14001 (EMS)		ISO 9001 (QMS)		Proportion of ISO 14001 certificates issued and withdrawn	Proportion of ISO 9001 certificates issued and withdrawn
		Number of certificates from 2003 to 2012	Number of withdrawn certificates	Number of certificates from 2003 to 2012	Number of withdrawn certificates		
1	Brazil	22,010	-640	149,351	-3,015	0.03	0.02
2	Netherlands	13,630	-224	122,882	-1,194	0.02	0.01
3	Nigeria	220	0	1,030	-6	0.00	0.01
4	South Africa	6,445	-232	32,493	-1,234	0.04	0.04
5	China	413,818	-19,142	2,187,470	-130,069	0.05	0.06
6	France	44,268	-800	240,333	-4,711	0.02	0.02
7	Italy	118,464	-1,466	1,135,569	-9,271	0.01	0.01
8	UK	96,989	-2,232	433,813	-10,802	0.02	0.02
9	US	49,682	-877	331,348	-5,061	0.02	0.02

Adapted from ISO (2012).

Conversely, it is possible to integrate the two process standards (as IMS or MMS) as scholars have established (see Low and Tan, 2005; Zeng et al., 2005). Secondly, Neumayer and Perkins (2003) identified supply-side factors as being responsible for the uneven adoption of ISO 14001 at the global level. They found a significant negative correlation between productivity and certification counts and a positive correlation between income and certification counts. In the US and UK, Morrow and Rondinelli's (2002) study found parent corporation's requirement to improve environmental performance, preference for getting beyond regulatory compliance, and finding cost-cutting opportunities as motivators to adopting EMS.

Furthermore, they found developing competitive advantage as motivating the adoption of EMS among companies operating in international market. On a larger scale, Raines' (2002) study found EMS motivation as varying among companies in developed and developing economies. In the latter, plain enthusiasm to "be a good neighbor" as well as being able "to raise employees' level of environmental awareness" outweigh profit-related motives.

From ISO (2012), China, Italy and Spain have been the three top leading countries in the number of ISO 9001 certificates while China, Japan and Italy have been the three top leading countries in the number ISO 14001 certificates. Among the four countries, save for Spain and Japan that are not among Nigeria's top trading partners as identified earlier in this section, there is a dearth of existing literatures documenting Italy's FDI in Nigeria.

On the other hand, in spite of China's role in infrastructure development in Nigeria (see Babatunde and Low, 2013a), it has, nonetheless, come under intense scrutiny in Nigeria. Claims against China's operations in Nigeria include its, supposedly, being exploitative in its practices (see CCS, 2013b; 2013c), bringing in its own suppliers and being the infamous world factory for poor quality services (see CCS, 2013d).

With culture being ingrained in values held by different countries thus manifesting in different practices among countries (see Ochieng and Price, 2009) the role of culture on China-Nigeria relationship alike has been theorized (see Babatunde and Low, 2008) and proven, empirically (see Babatunde, 2013).

In response to the different accusations on its FDI operations, in 2011, China published its first official white paper on its foreign aid (see Information Office of the State Council, 2011 for detailed paper). *Inter alia*, the paper concluded identifying global challenges such as climate change, food crisis, energy and resource security and epidemic of diseases confronting the developing economies as well as worsening global economy imbalance.

China posited it purposes to tackle the challenges by working together with other countries (developed and developing) to create a prosperous and harmonious world with lasting peace in support of the UN MDGs. Correspondingly, China's aids to Nigeria has since been multifaceted as presented in Appendix II, which shows China's past five years engagement with Nigeria. More importantly, it nullifies claims of China's focus on natural resources (for example, see Butts and Bankus, 2009; Taylor, 2006). Conversely, there are notable studies on policies guiding China's FDI (see Long, 2006) and its significant role as an official development agent in developing economies (see UNDP, 2011: 145-83).

5.4. Sustainability as an Inside-Out Approach: Lessons from China

Scholars have investigated the relevance of ancient wisdom to modern day thinking and, by extension, in resolving the challenges the world face today (see Avegalio, 2009; Van Binsbergen, 2009) including sustainability (see Appiah-Opoku, 2006; Malunga and Banda, 2004). Thusly, there is a Chinese proverb that "One generation plants a tree, another gets the shade". Remarkably, Chinese traditional concept of *Feng-shui*, based on the philosophy of harmonizing people and nature, has been found relevant to sustainability (see Chen and Nakama, 2004). Overseas, China's dominance in quality certifications coupled with the claim against it that it exports its suppliers from home justifies that Chinese firms in Nigeria have the requisite knowledge of sustainability. Cushing et al.'s (2004) study supports the foregoing position by concluding widespread adoption of EMS has increased environmental awareness among Chinese enterprises. Similarly, the English proverb that "When in Rome, do as the Romans do" accords well with China's operations in Africa. Zadek et al.'s (2009) study on Chinese firms in Africa has found most of the Chinese firms adopting CSR and, traditionally, relying on philanthropy and local laws; and less on ISO 14001 *per se* due to the more unfamiliar and/or controversial standards from Africa's multi-stakeholders. Correspondingly, China has been involved in similar socially responsible activities in Nigeria as can be deduced from Appendix II.

From Chan and Hu (2003), China is probably one of the fastest urbanizing countries in the world whose urbanization, in the late 1990s, arose majorly from *rural-urban migration* unlike other developing countries, where *urban natural increase* has been a significant source of urban demographic growth.

While Chan and Hu's (2003) conclusion can be attributed to China's one-child policy, it, nonetheless, agrees with Gaston's (2010: 11) position that *rural-urban migration* accounts for a substantial proportion of urbanization among developing economies.

Perhaps, what is unique to China is its urban design, which harmonizes with the Chinese context because of its sheer scale as well as extent and characteristics of urbanization in recent years (see Chen and Thwaites, 2013). Chen and Thwaites (2013: 2) posited that in spite of dramatic change in urban form, Chinese cities have maintained a remarkable cultural continuity over thousands of years to a great extent that its exploration can elicit critical lessons for other countries.

Critically, Chen and Thwaites (2013: 2) noted the scale and development in city expansion worldwide reflects processes of manufacture rather than growth and might well lead to urban populations whose lives are at odds with, and, at extremes, distorted by the urban fabric they inhabit.

China's contrast, based on typomorphology, has seen to the emerging new field of *Chinese urban design* (as against urban design in China) (see Chen and Thwaites, 2013 for details).

China's foregoing experience is not atypical of African countries. The difference lies in a self-explanatory Nigerian proverb that "Wealth diminishes with usage; learning increases with use" (see Mwiti, 2012). Africa's resource management problems are festering on the abandoning of indigenous resource management systems in favor of Western ones (see Konadu-Agyemang and Panford, 2006).

On the contrary to Western science, indigenous knowledge reconnects human beings to nature and respects culture (see Appiah-Opoku, 2006). In Nigeria, indigenous architecture has been found to be sustainable; however, under-developed and, as a result, not sustaining housing (see Odebiyi, 2010). Nigeria's sustainable housing challenge hinges on "knowledge gap" and not "lack of capacity" that developing the local capacity in all aspects and complementing with the latest knowledge would suffice (see Akeju, 2007). A pragmatic and efficient solution to Nigeria's sustainability and urban development debacle is to tackle poverty and environmental degradation (see Lawanson, 2005). Undoubtedly, [single-handedly] achieving sustainable urbanization would be herculean in developing economies (see Cohen, 2005). More so, there are inherent limitations of indigenous knowledge (see Briggs, 2005) and Western knowledge alike (see Hoppers, 2002: 8); thus necessitating a hybrid system that responds, complementarily, to environmental concerns (see Mercer et al., 2010).

CONCLUSION

This study's first objective was to investigate the three key issues of public housing policy in Nigeria.

Contrastingly, it was found that public housing constitutes about 10 percent of the existing housing stock in Nigeria with the significant portion of the private housing unqualified as quality housing. Summarily, prevailing public housing policies lack the characteristics of being culturally-acceptable, livable, affordable, sociable, and scalable, construed as CLASS. It was also found that past efforts have not been holistic; thus, lack the impetus for action. The assessments have not been specific, measurable, attainable, realistic, and time-bound (SMART according to Doran, 1981).

Consequently, this study has distilled a SMART CLASS alternative, which lends credence to Nigeria's endogenous and exogenous factors alike (see Section 4). It is recommended that subsequent housing policies in Nigeria elicit lessons from the alternative and apply the alternative for further validation. The second objective was to investigate the roles of major foreign players and how they can, uncompromisingly, complement sustainable urbanization in Nigeria. It was conceptualized that globalization influences urbanization and urbanization impacts on sustainable development (see Figure 1).

Consequently, it was established that Nigeria's prevailing close-loop urbanization model (see Figure 5) as typified by the concentration of developments in Lagos and Abuja is unsustainable. Firstly, the model is a pseudo measure because, at best, it shifts the trajectory of migration from one city to another to widening the rural-urban divide further. Secondly, FDI's concentration in capital cities and coercively acquired and selective rural areas has only served the interests of the foreign players. It is recommended that FDI should be decentralized down to Nigeria's local government areas and pre-assessed adopting the performance-based SD index developed in this study (see Section 2.2).

The last objective was to investigate environmental management system among foreign and local firms in Nigeria, identify the gap, and recommend solutions.

It was found, in addition to playing significant multifaceted role in Nigeria's development (see Appendix II), China is, notably, playing a domineering role in IMS certification incorporating the EMS alike. Based on cultural values and prevalence of EMS certification, Chinese firms possess the requisite environmental knowledge, hypothetically (see Section 5.4).

The continued influx of Chinese firms into Nigeria will, eventually, influence local firms' adoption of international standards, which is seriously lagging behind in comparison with its top trading partners' (see Table 7), and with that improve their competitiveness alike.

On the other hand, local firms can develop their indigenous knowledge further by bringing this to the fore in a hybrid system to complement advanced foreign technology (Western and Eastern).

It is also worth emphasizing effective domestic regulatory frameworks and pragmatic international trade agreements underpin the economic development (as opposed to mere economic growth taking a cue from Daly, 1990) that FDI purposes to achieve.

Urbanization, or more aptly, unsustainable urbanization has brought untold damages. Globalization is a tool through which a developing economy bridges the gap in its development as measured against that of the developed economy.

However, efficiency necessitates globalization and its associated advanced technology to be cognizant of the fact that the world is not monolithic.

There are realities that exist along the four aspects of SD between and within the developed and developing economies. As a whole, the world's resources are finite and so are human's capacities to intervene with world's resources.

Unwholesome economic development is menacingly changing climate and impacting on the world's erstwhile rich biodiversity. On a final note, this study also underscores the importance of culture, noting culture as the foundation on which politics and government that determines economic systems rest.

Culture adopts technology as a tool to propagate itself, which, in consequence, means that culture affects technology adoption and innovation that follows. Similarly, culture in the form of certain dominant values (for example, collective versus individual benefits, long-term versus short-term focus) shapes action and development that follows.

As such, culture pervades acceptance of responsibilities to emphasize sustainability of natural resources, encourage and foster capacity building, and inspire proactive environmentally sensitive development. Urbanization dates back the 18th century but has become an ingrained characteristic of a thriving economy that it is difficult to tame. It then follows that the decisions being taken to meet the present needs must complement the ability of the future generations to meet their own needs.

Appendix I. Nigeria: States, Administrative Units, Areas, Population and Growth

No	Administrative Units				Area	Population		
	State	Capital	Local Government Area (LGA)	Former LGAs or LGAs now part of another province	Area of the states (in sq.km.)	Population (1991 census)	Population (2011 projection)	Population Growth Rate (in %)
1	Abia	Umuahia	17	5	4,902.24	2,329,487	3,250,816	39.55
2	Adamawa	Yola	21	Nil	38,823.31	2,102,053	3,667,419	74.47
3	Akwa Ibom	Uyo	31	3	6,772.09	2,409,613	4,612,067	91.40
4	Anambra	Awka	21	2	4,816.21	2,796,475	4,796,408	71.52
5	Bauchi	Bauchi	20	9	49,933.87	4,351,007	5,499,737	26.40
6	Bayelsa	Yenagoa	8	Nil	9,415.76	0	1,966,426	0
7	Benue	Makurdi	23	Nil	31,276.71	2,753,079	4,931,135	79.11
8	Borno	Maiduguri	27	Nil	75,480.91	2,536,003	4,930,077	94.40
9	Cross River	Calabar	18	2	21,636.60	1,911,279	3,337,517	74.62
10	Delta	Asaba	25	1	17,239.24	2,590,491	4,813,917	85.83
11	Ebonyi	Abakaliki	13	Nil	6,421.23	0	2,499,272	0
12	Edo	Benin City	18	Nil	19,819.28	2,172,005	3,694,087	70.08
13	Ekiti	Ado-Ekiti	16	Nil	5,887.89	0	2,794,575	0
14	Enugu	Enugu	17	7	7,660.17	3,154,380	3,788,319	20.10
15	Gombe	Gombe	11	Nil	17,982.03	0	2,768,452	0
16	Imo	Owerri	27	2	5,182.82	2,485,635	4,597,500	84.96
17	Jigawa	Dutse	27	Nil	24,515.62	2,873,121	5,031,103	75.11
18	Kaduna	Kaduna	23	2	45,711.19	3,935,618	7,087,226	80.08
19	Kano	Kano	44	Nil	21,276.87	5,810,470	11,058,314	90.32
20	Katsina	Katsina	34	Nil	24,971.22	3,753,133	6,725,626	79.20
21	Kebbi	Birnin Kebbi	21	Nil	37,727.97	2,068,490	3,793,585	83.40
22	Kogi	Lokoja	21	2	29,581.89	2,147,756	3,841,884	78.88

Appendix I. (Continued)

No	Administrative Units				Area	Population		
	State	Capital	Local Government Area (LGA)	Former LGAs or LGAs now part of another province	Area of the states (in sq.km.)	Population (1991 census)	Population (2011 projection)	Population Growth Rate (in %)
23	Kwara	Ilorin	16	Nil	34,467.54	1,548,412	2,742,093	77.09
24	Lagos	Ikeja	20	Nil	3,496.45	5,725,116	10,668,139	86.34
25	Nasarawa	Lafia	13	Nil	27,271.50	0	2,167,120	0
26	Niger	Minna	25	Nil	74,108.58	2,421,563	4,674,381	93.03
27	Ogun	Abeokuta	20	Nil	16,980.55	2,333,726	4,412,299	89.07
28	Ondo	Akure	18	17	15,195.18	3,784,902	4,012,105	6.00
29	Osun	Oshogbo	30	3	8,699.84	2,158,143	3,999,800	85.34
30	Oyo	Ibadan	33	4	28,245.26	3,452,720	6,596,392	91.05
31	Plateau	Jos	17	10	27,216.95	3,318,257	3,663,428	10.40
32	Rivers	Port Harcourt	23	9	10,432.28	4,309,557	6,144,673	42.58
33	Sokota	Sokoto	23	12	33,776.89	4,470,176	4,292,416	-3.98
34	Taraba	Jalingo	17	Nil	60,291.82	1,512,184	2,647,413	75.07
35	Yobe	Damaturu	17	Nil	46,909.76	1,399,687	2,757,022	96.97
36	Zamfara	Gusau	14	Nil	35,170.63	0	3,838,160	0
37	Abuja FCT	Abuja	6	Nil	7,753.85	371,674	2,193,613	490.20
Total			775		937,052.16	88,9[86,212]	164,294,516	

Adapted from GeoHive (www.geohive.com/cntry/nigeria.aspx, last accessed November 25, 2013). Please note: Area is about 923,769 sq.km (see CIA, 2013; NBS, 2010: 3).

Appendix II. Some Chinese Investments in Nigeria Between 2009 and 2013

No.	Project	Value in USD	Parties Involved	CCS Report Date
1	China's Huawei offers ICT training program to 1000 Nigerian women	NA	Huawei and Nigerian Federal Ministry of Telecommunication Technology (FCMT)	Nov. 8, 2013
2	An agreement to construct the Zungeru Hydroelectric Power Project	1.3 billion	China EXIM Bank, Federal Government of Nigeria (FGN), China National Electric and Engineering Corporation and SinoHydro	Oct. 4, 2013
3	Huawei aims to double smartphone sales to 200,000 in Nigeria in 2013	NA	Huawei and Galaxy Backbone (Nigeria)	Sept. 27, 2013
4	Nigerian company announces keel laying of the first of two platform supply vessels	NA	Pheranzy Offshore and Hongda Marine	Sept. 13, 2013
5	Nigeria and China Sign MoU for Power grid Projects	500 million	China EXIM Bank, FGN and Xian Electric (China)	30 Aug., 2013
6	Chinese firm lays foundation stone for major light rail project in Nigeria (Abuja)	500 million	China EXIM Bank, FGN and China Civil Engineering Construction Corporation (CCECC)	26 July, 2013
7	Chinese firm proposes 20,000MW power plant in Nigeria	3.7 billion	NA	May 10, 2013
8	Nigerian company signs a USD1.25 billion deal with Chinese firms	1.25 billion	Globacom Nigeria Ltd, Huawei and ZTE	May 4, 2013
9	China Assisted Hospital [150-bed] Commissioned in Nigeria	12.5 million	China EXIM Bank	Feb. 1, 2013
10	Nigeria seeks USD1.5 billion from China for Rice and Cassava	1.5 billion	China EXIM Bank and Nigerian Ministry of Agriculture and Rural Development	Dec. 14, 2012
11	China Buys a Stake in Nigeria's Oil Industry	2.46 billion	China Petrochemical Corporation and Total ADR	Nov. 23, 2012

Appendix II. (Continued)

No.	Project	Value in USD	Parties Involved	CCS Report Date
12	Chinese Firm Sets Up Railway Tech Training Center in Nigeria	NA	CCECC	Nov. 9, 2012
13	Chinese Firm to Build USD32 million Civic Centre and Museum in Southern Nigeria	32 million	Tianjin Yuyang Construction Company	Oct. 12, 2012
14	Chinese Company to Modernize Railway in Nigeria (Lagos-Ibadan Double Track)	1.48 billion	CCECC	Aug. 31, 2012
15	50 million investment in Lekki Free Trade Zone (Primed for USD 1 billion investment)	1 billion	China Railway Construction Corporation.	Aug. 10, 2012
16	Chinese and Nigerian Firms Plan Coal Power Plant	4 billion	Sepco III	July 20, 2012
17	Chinese Satellite Said to Create 150,000 Jobs for Nigerians	95 million savings per year	China Great Wall Industry Corporation	Mar. 23, 2012
18	Nigerian Finance Minister to Meet Chinese Bank Officials Over USD 3 billion Loans	3 billion	FGN	Mar. 2, 2012
19	Sinoma Completes USD1 billion Cement Project in Nigeria	1 billion	Sinoma and Dangote Group Nigeria	Feb. 17, 2012
20	China, Nigeria Spend USD 629 million on construction of Abuja to Kaduna Rail	850 million	China EXIM Bank, FGN and CCECC	Jul. 29, 2011
21	Nigeria, China Sign Kainji Hydro Plant Modernization Contract	82 million	Power Holding Company of Nigeria, Sinohydro Corporation (China) and Harbin Electricity Corporation of China	Jul. 15, 2011
22	Nigeria-China Trade Hits USD 7.76 billion	7.76 billion		Apr. 8, 2011

No.	Project	Value in USD	Parties Involved	CCS Report Date
23	China Merchants to Buy Stake in Nigeria Port Operator	154 million	China Merchants Holdings International and China-Africa Development Fund	Nov. 12, 2010
24	China Plans USD 20 billion Investments in Nigeria's Infrastructure	20 billion	Nigerian Investment Promotion Commission	Sept. 23, 2010
25	China and Nigeria to Build Free Trade Zone in Lagos		Chinese Government, Lagos State Government (Nigeria) and Lekki Free Trade Zone Development Company	Sept. 10, 2010
26	China's Sinopec Reports Oil Discovery in Nigeria			Jul. 16, 2010
27	Nigeria and China Sign Agreement to Build USD 8 billion Refinery	8 billion	China State Construction Engineering Corp and National Petroleum Corp (Nigeria).	Jul. 9, 2010
28	Nigeria and China: The Sunshine Deal	23 billion	Nigerian National Petroleum Corporation and China State Construction Engineering Corporation	May 21, 2010
29	Nigerian and Chinese Governments to Build Rural Schools	1.1 million	FGN and the Chinese government	Aug. 7, 2009
30	Nigeria, China Hold Trade Forum to Boost Commercial Links	NA	FGN and the Chinese government	Jul. 17, 2009
31	Nigeria's First Bank Gains Approval for China Office	NA	First Bank Nigeria Plc	Jul. 10, 2009
32	China Makes Dramatic Entry into Nigerian Oil Sector	7.24 billion	FGN and the Chinese government	Jul. 3, 2009
33	China, Nigeria State Firms Find Oil in Niger Delta		SIPEC and Nigerian Petroleum Development Company	Jun. 19, 2009
34	Nigeria Expects More Investment from China		FGN and the Chinese government	Mar. 20, 2009
35	Nigeria, Chinese Firms Sign USD 16 million Pact on Vehicle Plant	16 million	NA	Mar. 13, 2009

Adapted from CCS (www.ccs.org.za/?cat=9andpaged=2, last accessed November 28, 2013).

Note: All electronic media were last accessed on November 28, 2013.

REFERENCES

Abasiekong, O. (2013). World Bank Decries Poor Quality of Statistics and Data in Nigeria. *Business News* (January 2). http://businessnews.com.ng/2013/01/02/world-bank-decries-poor-quality-of-statistics-and-data-in-nigeria/.

Abegunrin, O. (2006). The Military and Nigerian Political Economy in the Global System. In: K. Konadu-Agyemang and K. Panford (Eds.), *Africa's Development in the Twenty-first Century: Pertinent Socio-Economic and Development Issues* (pp. 235-248). Hampshire, England: Ashgate Publishing Limited.

Achankeng, E. (2003). Globalization, Urbanization and Municipal Sold Waste Management in Africa. In: *Proceedings of the African Studies Association of Australasia and the Pacific 26th Annual Conference – African on a Global Stage.*

Adams, O. (1997). Contractor Development in Nigeria: Perceptions of Contractors and Professionals. *Construction Management and Economics*, 15(1), 95-108.

Adeogun, O. B. and Taiwo, A. A. (2011). Housing Delivery Through Public-Private Partnership in Nigeria and the Case for Beneficiaries' Involvement. *Journal of Construction Project Management and Innovation*, 1(2), 63-79.

Aderinokun, K. and Eze, C. (2013). Chinese Investors Besiege Nigeria as Annual Trade Hits $13bn. *This Day Live* (July 12). www.thisdaylive.com/articles/chinese-investors-besiege-nigeria-as-annual-trade-hits-13bn/153107/.

Aduwo, E. B., Ibem, E. O. and Opoko, A. P. (2013). Residents' Transformation of Dwelling Units in Public Housing Estates in Lagos, Nigeria: Implications for Policy and Practice. *International Journal of Education and Research*, 1(4).

AfDB (African Development Bank) and OECD (Organization for Economic Co-operation and Development) (2008). *African Economic Outlook 2008* (Nigeria). Paris, Francis: OECD.

AfDB, OECD, UNDP, and UNECA (2011). *African Economic Outlook: Nigeria*. Issy Les Moulineaux, France: OECD Development Centre.

Agbedo, O. (2013). Dry Construction Can Address Nigeria's Housing Deficit. *Guardian News* (August 2). www.statewww.ngrguardiannews.com/business-news/128840-dry-construction-can-address-nigerias-housing-deficit.

Agtmael, A. W. (2007). *The Emerging Markets Century: How a New Breed of World-Class Companies is Overtaking the World*. New York: Free Press.

Ahlstrom, D. and Bruton, G. D. (2010). *International Management: Strategy and Culture in the Emerging World*. US: Southwestern Cengage Learning.

Aid Data (2009). Housing Project and Greater Abuja Water Project. *Aid Data* (ID 2138). http://china.aiddata.org/projects/2138.

Aid Data (2013). $6 Billion Towards Nigeria's Housing Sector. *Aid Data* (ID 30454). http://china.aiddata.org/projects/30454.

Akeju, A. A. (2007). Challenges to Providing Affordable Housing in Nigeria. *A paper presented at the 2nd Emerging Urban Africa International Conference on Housing Fincnce in Nigeria*. Abuja, Nigeria: Sehu Yar' adua Center (October 17-19).

Akinsuyi, Y. and Sanni, T. (2013). 31 States to Experience Devastating Flood in 2013, Says Report. *This Day Live* (April 27). www.thisdaylive.com/articles/31-states-to-experience-devastating-flood-in-2013-says-report/146070/.

Al-Rodhan, N. R. F. and Stoudmann, G. (2006). *Definitions of Globalization: Comprehensive Overview and Proposed Definition*. Switzerland, Geneva: Center for Security Policy.

Alagbe, W. (2013). Exploring Indigenous Strategies for Affordable Housing Delivery for the Urban Poor. *A presentation at the 3rd ASO National Housing Exhibition and Conference: Achieving Affordable Housing Delivery by Creating 500,000 Housing Units by 2016*. Abuja, Nigeria: International Conference Center (April 11-13).

Alston, M. and Bowles, W. (2003). *Research for Social Workers: An Introduction to Methods* (2nd edition). New South Wales, Australia: Allen and Unwin.

Alufohai, A. J. (2013). The Lagos State 2010 Mortgage Law and the Supply of Housing. *A presentation at the FIG Working Week 2013: Environment for Sustainability*. Abuja, Nigeria: International Conference Center and NICON Luxury Hotel (May 6-10).

Aluko, O. (2012). The Effects of Land Use Act on Sustainable Housing Provision in Nigeria: The Lagos State Experience. *Journal of Sustainable Development*, 5(1), 114-122.

Aniekwu, A. (1995). The Business Environment of the Construction Industry in Nigeria. *Construction Management and Economics*, 13(6), 445-455.

Appiah-Opoku, S. (2006). Indigenous Knowledge and Environmental Management in Africa: Evidence from Ghana. In: K. Konadu-Agyemang and K. Panford (Eds.), *Africa's Development in the Twenty-first Century: Pertinent Socio-Economic and Development Issues* (pp. 205-216). Hampshire, England: Ashgate Publishing Limited.

Aribigbola, A. (2013). Cause and Consequences of Housing Policy Failure in Nigeria. *British Journal of Social Sciences*, 1(6), 18-31.

Ashkin, R. (2013). Innovative Building Technologies: The Social Housing Angle. *A Presentation at Housing Africa 2013*. Abuja, Nigeria (August 14).

Asiedu, E. (2002). On the Determinants of Foreign Direct Investment to Developing Countries: Is Africa Different? *World Development*, 30(1), 107-119.

Ayozie, D. O. (2011). The Role of Small Scale Industry in National Development in Nigeria. *Universal Journal of Management and Social Sciences*, 1(1), 23-41.

Babatunde, O. K. (2013a). Cross Cultural Management of Quality Performance of Chinese Construction Firms in Nigeria. *PhD Thesis submitted to the Department of Building, National University of Singapore* (October).

Babatunde, O. K. and Low, S. P. (2008). Cross-Cultural Crisis Management on Quality: A Case Study of Chinese Contractors in Nigeria. In: C. C. Feng et al. (Eds.), *Proceedings of the CRIOCM on Advancement of Construction Management and Real Estate* (pp. 347-352). Beijing, China: Renmin University (October 31-November 3).

Babatunde, O. K. and Low, S. P. (2013a). Chinese Construction Firms in the Nigerian Construction Industry. *Habitat International*, 40, 18-24.

Babatunde, O. K. and Low, S. P. (2013b). Transition from ISO 9001 to TQM: Recommendations for the Nigerian Construction Industry. *Malaysian Construction Research Journal*, 12 [In Press].

Briggs, J. (2005). The Use of Indigenous Knowledge in Development: Problems and Challenges. *Progress in Development Studies*, 5(2), 99-114.

Brock, J. (2013). Nigeria's Property Boom: Only for the Brave. *Reuters* (September 15). www.reuters.com/article/2013/09/15/us-nigeria-property-analysis-idUSBRE98E03620130915.

Brown, M., Hall, C. A. S. and Wackernagel, M. (2000). Comparative Estimates of Sustainability: Economic, Resource Base, Ecological Footprint, and Emergy. In: C. A. S. Hall, C. L. Perez and G. Leclerc

(Eds.), *Quantifying Sustainable Development* (pp. 695-714). San Diego, US: Academic Press.

Brundtland, G. H. (1987). *Report of the World Commission on Environment and Development: Our Common Future*. Oslo: United Nations.

BSI (British Standards Institution) (2013). *A Comparison between ISO 14001 Environmental Management and ISO 9001 Quality Management*. Minton Keynes, United Kingdom: BSI. www.bsigroup.ae/upload/MEA/Performance/ISO14001_vs_ISO9001.pdf.

Business Day (2013). Rural Infrastructure: The Lagos Example. *Business Day* (February 1). www.businessdayonline.com/NG/index.php/analysis/editorial/51001-rural-infrastructure--the-lagos-example.

Büthe, T. and Milner, H. V. (2008). The Politics of Foreign Direct Investment into Developing Countries: Increasing FDI through International Trade Agreements? *American Journal of Political Science*, 52(4), 741-762.

Butts, K. H. and Bankus, B. (2009). China's Pursuit of Africa's Natural Resources. *Center for Strategic Leadership and Development, US Army War College* (Volume 1-09).

CASI (Communities and Social Identities) (2012). *A Brief History of Sustainable Development*. Bath Spa University: CASI Research Group.

CCS (Centre for Chinese Studies) (2009a). Oil-Rich Angola Has Become China's Largest Trade Partner in Africa. *The Weekly China Briefing* (January 23). Stellenbosch, South Africa: CCS. www.ccs.org.za/?cat=9andpaged=2.

CCS (2009b). Angola Becomes China's Largest Trade Partner in Africa. *The Weekly China Briefing* (October 2). Stellenbosch, South Africa: CCS. www.ccs.org.za/?cat=9andpaged=2.

CCS (2012). China Starts Building 6 million Homes Next Year. *The Weekly China Briefing* (March 11). Stellenbosch, South Africa: CCS. www.ccs.org.za/?cat=9andpaged=2.

CCS (2013a). China to Provide Africa with US$ 1 Trillion Financing. *The Weekly China Briefing* (November 22). Stellenbosch, South Africa: CCS. www.ccs.org.za/?cat=9andpaged=2.

CCS (2013b). China-Africa Honeymoon Over, Claims Nigerian Central Bank Governor. *The Weekly China Briefing* (March 15). Stellenbosch, South Africa: CCS. www.ccs.org.za/?cat=9andpaged=2.

CCS (2013c). Nigerian Central Bank Governor's Comments Draws Chinese Response. *The Weekly China Briefing* (March 22). Stellenbosch, South Africa: CCS. www.ccs.org.za/?cat=9andpaged=2.

CCS (2013d). Nigerian Workers Protest Conditions at Chinese Construction Projects. *The Weekly China Briefing* (February 8). Stellenbosch, South Africa: CCS. www.ccs.org.za/?cat=9andpaged=2.

Chan, K. W. and Hu, Y. (2003). Urbanization in China in the 1990s: New Definition, Different Series, and Revised Trends. *The China Review*, 3(2), 49-71.

Chen, B. X. and Nakama, Y. (2004). A Summary of Research History on Chinese Feng-shui and Application of Feng-shui Principles to Environmental Issues. *Kyushu Journal of Forest Research*, 57(3), 297-301.

Chen, F. and Thwaites, K. (2013). *Chinese Urban Design: The Typo-morphological Approach*. Surrey, England: Ashgate.

Chinese Embassy (2004). China-Nigeria Relations. Embassy of the People's Republic of China in the Federal Republic of Nigeria (*Bilateral Relations*). http://ng.china-embassy.org/eng/zngx/t142490.htm.

CIA (Central Intelligence Agency) (2013). *The World Factbook - Nigeria*. Washington, D.C.: CIA. https://www.cia.gov/library/publications/the-world-factbook/geos/ni.html.

CIL (Centre for International Learning) (2011). Culture Does Not Exist. *Magazine Intercultures*, 2(3). www.international.gc.ca/cfsi-icse/cil-cai/magazine/v02n03/1-3-eng.asp.

Cohen, B. (2005). Urbanization in Developing Countries: Current Trends, Future Projections, and the Key Challenges for Sustainability. *Technology in Society*, 28, 63-80.

Coker, A. O., Awokola, O. S., Omolaiye, P. O., and Booth, C. A. (2007). Challenges of Urban Housing Quality and Its Associations with Neighborhood Environments: Insights and Experiences of Ibadan City, Nigeria. *Journal of Environmental Health Research*, 7(1), 21-30.

Cooper, J. (2006). Of BRICs and Brains: Comparing Russia with China, India, and Other Populous Emerging Economies. *Eurasian Geography and Economics*, 47(3), 255-284.

Corporate Guides (2010). *FDI Overview*. www.corporate-nigeria.com/index/fdi/foreign_direct_investment_overview.html.

Crawford, L. (2007). Global Body of Project Management Knowledge and Standards. In: P. W. G. Morris and J. K. Pinto (Eds.), *The Wiley Guide to Project Organization and Project Management Competencies* (pp. 206-252).

Cushing, K. K., McGray, H. and Lu, H. (2005). Understanding ISO 14001 Adoption and Implementation in China. *International Journal of Environment and Sustainable Development*, 4(3), 246-268.

Daly, H. E. (1990). Toward Some Operational Principles of Sustainable Development. *Ecological Economics*, 2(1), 1-6.

Davis, K. (1955). The Origin and Growth of Urbanization in the World. *American Journal of Sociology*, 60(5), 429-437.

Delmas, M. (2001). Stakeholders and Competitive advantage: The Case of ISO 14001. *Production and Operations Management*, 10(3), 343-358.

Dikko, H. A. (2002). Cost Control Models for Housing and Infrastructure Development. *A presentation at FIG XXII International Congress: Affordable Housing*. Washington D.C., US (April 19-26).

Dingwerth, K. and Pattberg, P. (2006). Global Governance as a Perspective on World Politics. *Global Governance*, 12(2), 185-203.

Dodo, B. D. (2004). Why Past Housing Schemes Failed. *Daily Trust* (March 10). http://allafrica.com/stories/200403100677.html.

Doran, G. T. (1981). There's a S.M.A.R.T. Way to Write Management's Goals and Objectives. *Management Review* (AMA Forum, November), 35-36.

Drexhage, J. and Murphy, D. (2010). Sustainable Development: From Brundtland to Rio 2012. *Background Paper prepared for the High Level Panel on Global Sustainability*. New York: United Nations Headquarters (September). www.trademarksa.org/node/3423.

Egene, G. (2013). Expect Cheaper Mortgages in 2014. *This Day Live* (July 24). www.thisdaylive.com/articles/okonjo-iweala-expect-cheaper-mortgages-in-2014/154193/.

Eriksson, C. (2013). *Economic Growth and the Environment*. Oxford, UK: Oxford University Press.

Fashoyin, T. (1996). The Effects of Adjustment Measures on Nigerian Labor. In: A. Y. Yansané (Ed.), *Prospects for Recovery and Sustainable Development in Africa* (pp. 162-179). Westport, Connecticut, London: Greenwood Press.

Ferguson, M. (1992). The Mythology About Globalization. *European Journal of Communication*, 7(1), 69-93.

Filaba, M. A. (2008). Paradox of African Capital Cities: Case Study of Global Capitalism and Urban Poor in Lagos and Abuja Capital Cities in Nigeria. *African Research Review*, 1(1), 5-13.

Finkelstein, L. S. (1995). What Is Global Governance? *Global Governance*, 1 (3), 367-372.

Fox, T. J. and Wallach, D. (2013). *Green Town US: The Handbook for America's Sustainable Future*. United States: Hatherleigh Press.

Frost, J. (2013). How You Can Influence Next Edition of ISO 14001. Geneva, Switzerland: International Organization for Standardization (ISO). www.iso.org/iso/home/news_index/news_archive/news.htm?refid=Ref1713.

Gaillard, J. (2010). Measuring Research and Development in Developing Countries: Main Characteristics and Implications for the Frascati Manual. *Science, Technology and Society*, 15(1), 77-111.

Gaston, K. J. (Ed.) (2010). Urbanization. In: *Urban Ecology* (pp. 10-34). Cambridge, United Kingdom: Cambridge University Press.

GCP (Global Construction Perspectives) and OE (Oxford Economics) (2011). *Global Construction 2020: A Global Forecast for the Construction Industry over the next Decade to 2020*. London, UK: GCP and OE.

GeoHive (2013). Country Data: Nigeria. www.geohive.com/cntry/nigeria.aspx.

Gerring, J. (2004). What is a Case Study and What is it Good for? *American Political Review*, 98(2), 341-354.

Gibberd, J. (2002). The Sustainable Building Assessment Tool: Assessing how Buildings can Support Sustainability in Developing Countries. *Paper presented at the Built Environment Professions Convention*. Johannesburg, South Africa: Current Events (May 1-3).

Gibberd, J. (2008). Developing a Sustainable Development Approach for Buildings and Construction Processes. In: J. Yang, P. S. Brandon and A. C. Sidwell (Eds.), *Smart and Sustainable Built Environments* (pp. 300-311). Oxford, UK: Blackwell Publishing Ltd.

Giddings, S. W. (2007). *Housing Challenges and Opportunities in Sub-Saharan Africa*. Washington DC: International Housing Coalition.

Giffinger, R., Fertner, C., Kramar, H., Kalasek, R., Pichler-Milanović, N., and Meijers, E. (2007). *Smart Cities: Ranking of European Medium-Sized Cities*. Vienna: Centre of Regional Science.

Görg, H. and Greenaway, D. (2003). Much Ado About Nothing? Do Domestic Firms Really Benefit from Foreign Direct Investment? *IZA Discussion Paper Series* (No. 944). In: Bonn (Ed.), *Forschungsinstitut zur Zukunft der Arbeit.*

Hall, C. A. S. (2000a). The Changing Tropics. In: C. A. S. Hall, C. L. Perez and G. Leclerc (Eds.), *Quantifying Sustainable Development* (pp. 3-18). San Diego, US: Academic Press.

Hall, C. A. S. (2000b). The Theories and Myths That Have Guided Development. In: C. A. S. Hall, C. L. Perez and G. Leclerc (Eds.),

Quantifying Sustainable Development (pp. 45-89). San Diego, US: Academic Press.

Hall, C. A. S. (2000c). The Myth of Sustainable Development. In: C. A. S. Hall, C. L. Perez and G. Leclerc (Eds.), *Quantifying Sustainable Development* (pp. 715-731). San Diego, US: Academic Press.

Hall, E. (1976). *Beyond Culture*. Garden City, New York: Doubleday Anchor Books.

Hall, P. A. and Watt, F. M. (1989). Stem Cells: The Generation and Maintenance of Cellular Diversity. *Development*, 106(4), 619-633.

Harris, J. M. (2003). Sustainability and Sustainable Development. In the International Society for Ecological Economics *Internet Encyclopedia of Ecological Economics*.

Harris, R. W. and Chib, A. (2012). Finding a Path to Influencing Policy. In: R. W. Harris and A. Chib (Eds.), *Linking Research to Practice: Strengthening ICT for Development Research Capacity in Asia* (pp. 175-186). Singapore: ISEAS Publishing.

Harrison, L. E. and Huntington, S. P. (Eds.) (2000). *Culture Matters: How Values Shape Human Progress*. New York, US: Basic Books.

HDB (Housing and Development Board) (2013). Public Housing in Singapore. *HDB InfoWEB* (September 23). www.hdb.gov.sg/fi10/fi10320p.nsf/w/AboutUsPublicHousing.

He, S., Nakada, D. and Morrison, S. J. (2009). Mechanisms of Stem Cell Self-Renewal. *Annual Review of Cell and Developmental Biology*, 25, 377-406.

Hillier, B. (2009). Spatial Sustainability in Cities: Organic Patterns and Sustainable Forms. In: D. Koch, L. Marcus and J. Steen (Eds.), *Proceedings of the 7th International Space Syntax Symposium* (Keynote 1). Stockholm: KTH

Hofstede, G. H. (1980). *Culture's Consequences: International Differences in Work- related Values*. London: Sage Publications.

Hofstede, G. H. and Bond, M. H. (1988). The Confucius Connection: From Cultural Roots to Economic Growth. *Organizational Dynamics*, 16(4), 4-21.

Hofstede, G. H. and Hofstede, G. J. (2005). *Cultures and Organizations: Software of the Mind*, 2nd edition. New York: McGraw-Hill.

Hofstede, G. H., Hofstede, G. J. and Minkov, M. (2010). *Cultures and Organizations: Software of the Mind* (3rd edition). New York, US: McGraw-Hill.

Hoppers, C. A. O. (Ed.) (2002). *Indigenous Knowledge and the Integration of Knowledge Systems: Towards a Philosophy of Articulation*. Claremont, South Africa: New Africa Books (Pty) Ltd.

Hortensius, D. (2013). Integrated Management Systems. *ISO News* (February 18). www.iso.org/iso/home/news_index/news_archive/news.htm?refid=Ref1709.

Hoskisson, R. E., Eden, L., Lau, C. M., and Wright, M. (2000). Strategy in Emerging Economies. *Academy of Management Journal*, 41(3), 249-267.

Ibem, E. O., Anosike, M. N. and Azuh, D. E. (2011). Challenges in Public Housing Provision in the Post-Independence Era in Nigeria. *International Journal of Human Sciences*, 8(2), 421-443.

IBM Institute for Business Value (2013). *Analytics: A Blueprint for Value*. NY, US: IBM Corporation. www-935.ibm.com/services/us/gbs/thought leadership/ninelevers/.

IDMC (Internal Displacement Monitoring Centre) (2013). *Nigeria: Fragmented Response to Internal Displacement Amid Boko Haram Attacks and Flood Season*. Geneva, Switzerland: IDMC.

IEMA (Institute of Environmental Management and Assessment) (2013). *The Environmentalist* (October 30). www.environmentalistonline.com.

IHC (International Housing Coalition) (2009). *The Challenge of Urban World: An Opportunity for US Foreign Assistance*. Washington D.C., US: IHC.

IISD (International Institute for Sustainable Development) (2012). SD Timeline. www.iisd.org/pdf/2012/sd_timeline_2012.pdf.

Information Office of the State Council (2011). China's Foreign Aid. The People's Republic of China: Information Office of the State Council. http://english.gov.cn/official/2011-04/21/content_1849913.htm.

IRIN (Integrated Regional Information Network) (2013). Flood Test Nigeria's Preparedness. *IRIN News* (August 27). www.irinnews.org/report/98644/floods-test-nigeria-s-preparedness.

Isaacson, W. (2011). *Steve Jobs*. Avenue of the Americas, New York: Simon and Schuster.

Isimbabi, M. J. (1999). Private Investments and African Development: Illusions and Realities. In: V. U. James (Ed.), *Sustainable Development in Africa: Prospects for the 21st Century* (pp. 29-37). San Francisco: International Scholars Publications.

ISO (International Organization for Standardization) (2000), *The ISO Survey of ISO 9000 and 14000 Certificates – Ninth cycle*. Geneva, Switzerland: ISO Central Secretariat.

ISO (2004). *ISO 14001:2004 (en) Environmental Management Systems – Requirements with Guidance for Use* (ISO Online Browsing Platform). www.iso.org/obp/ui/#iso:std:iso:14001:ed-2:v1:en:fig:1.

ISO (2009). *Environmental Management: The ISO 14000 Family of International Standards*. Geneva, Switzerland: ISO Central Secretariat.

ISO (2010). ISO 26000 – Social Responsibility. *ISO Standards* (September 11). www.iso.org/iso/home/standards/iso26000.htm.

ISO (2012). *ISO Survey 2012. ISO Standards* (December 12). www.iso.org/iso/home/standards/certification/iso-survey.htm.

James, V. U. (Ed.) (1999). Operationalizing Concepts of Sustainable Development in Africa. In: *Sustainable Development in Africa: Prospects for the 21st Century* (pp. 1-7). San Francisco: International Scholars Publications.

Jiboye, A. D. (2011a). Achieving Sustainable Housing Development in Nigeria: A Critical Challenge to Governance. *International Journal of Humanities and Social Science*, 1(9), 121-127.

Jiboye, A. D. (2011b). Urbanization Challenges and Housing Delivery in Nigeria: The Need for an Effective Policy Framework for Sustainable Development. *International Review of Social Sciences and Humanities*, 2 (1), 176-185.

Jibril, I. U. and Garba, K. T. (2012). The Challenges of Housing Development and Needs in Abuja Nigeria. *FIG Working Week 2012: Knowing to Manage the Territory, Protect the Environment, Evaluate the Cultural Heritage*. Rome, Italy (May 6-10).

Johnson, S. (2011). Marrying Tech and Art. *The Wall Street Journal* (August 27). http://online.wsj.com/article/SB10001424053111904875404576532342684923826.html.

Kaasa, A. and Vadi, M. (2008). *How Does Culture Contribute to Innovation? Evidence from European Countries*. Tartu, Estonia: Tartu University Press.

Kabir, B. and Bustani, S. A. (2009). A Review of Housing Delivery Efforts in Nigeria. *A Presentation at the International Sociological Association Research Committee 43 Workshop*. Glasgow: University of Glasgow (September).

Kavaliku, L. (2005). Culture and Sustainable Development in the Pacific. In: A. Hooper (Ed.), *Culture and Sustainable Development in the Pacific* (pp. 22-31). Canberra, Australia: Asia Pacific Press at the Australian National University.

Khanna, T. and Palepu, K. G. (2010). *Winning in Emerging Markets: A Road Map for Strategy and Execution.* Boston, Massachusetts: Harvard Business Press.

Kirkpatrick, C., Parker, D. and Zhang, Y. F. (2006). Foreign Direct Investment in Infrastructure in Developing Countries: Does Regulation make a Difference? *Transnational Corporations*, 15(1), 143-171.

Kolawole, Y. (2013). FHA Acquires 4,106 Hectares in 19 States. *Vanguard*, January 13. www.vanguardngr.com/2013/01/housing-fha-acquires-4106-hectares-in-19-states/?utm_source=dlvr.itandutm_medium=twitter.

Konadu-Agyemang, K. and Panford, K. (Eds.) (2006). Introduction and Book Overview. In: *Africa's Development in the Twenty-first Century: Pertinent Socio-Economic and Development Issues* (pp. 1-10). Hampshire, England: Ashgate Publishing Limited.

Konar, A. (2000). *Artificial Intelligence and Soft Computing: Behavioral and Cognitive Modeling of the Human Brain.* Boca Raton, London: CRC Press.

Kopytoff, I. (1986). The Cultural Biography of Things: Commoditization as Process. In: A. Appadurai (Ed.), *The Social Life of Things: Commodities in Cultural Perspective* (pp. 65-91). Cambridge, United Kingdom: Cambridge University Press.

Kroeber, A. L. and Kluckhohn, C. (1952). *Culture: A Critical Review of Concepts and Definitions.* New York: Vintage Books.

Lawanson, T. O. (2005). Challenges of Sustainability and Urban Development in Nigeria: Reviewing the Millennium Development Goals. In: W. Fadare et al. (Eds.), *Globalization, Culture and the Nigerian Built Environment, 2.* Osun State, Nigeria: Obafemi Awolowo University.

Lee, S. E. (2012). Sustainable Development through the adoption of Total Building Performance approach. *Presented at the Singapore Contractors Association Limited (SCAL) Environment Sustainability Conference.* Singapore: Gardens by the Bay (October 16).

Lee, S. G., Trimi, S. and Kim, C. S. (2013). The Impact of Cultural Differences on Technology Adoption. *Journal of World Business*, 48, 20-29.

Leggett, J. A. and Carter, N. T. (2012). Rio+20: The United Nations Conference on Sustainable Development. *Congressional Research Service (CRS) Report for Congress* (R42573, June 18). www.fas.org/sgp/crs/row/R42573.pdf.

Lehrer, J. (2011). Steve Jobs: "Technology Alone Is Not Enough". *The New Yorker*, October 7). www.newyorker.com/online/blogs/newsdesk/2011/10/steve-jobs-pixar.html.

Levy, Y. and Ellis, T. J. (2006). A Systems Approach to Conduct an Effective Literature Review in Support of Information Systems Research. *Informing Science: International Journal of an Emerging Transcdiscipline*, 9, 181-212.

Long, G. (2005). China's Policies on FDI: Review and Evaluation. In: T. H. Moran, E. M. Graham and M. Blomström (Eds.). *Does Foreign Direct Investment Promote Development?* (pp. 315-336) Washington DC, US: Institute for International Economics.

Low, S. P. and Tan, J. H. (2005). Integrating ISO 901 Quality Management System and ISO 14001 Environmental Management System for Contractors. *Journal of Construction Engineering and Management*, 131 (11), 1241-1244.

Lowe, S. (2011). *The Housing Debate*. Bristol, Great Britain: The Policy Press.

Lubbers, R. and Koorevaar, J. (2000). Primary Globalization, Secondary Globalization, and the Sustainable Development Paradigm – Opposing Forces in the 21st Century. *Expo 2000, OECD Forum for the Future: Conference on 21st Century Social Dynamics: Towards the Creative Society*. Berlin (Dec. 6-7, 1999).

Luthans, F. and Doh, J. P. (2012). *International Management: Culture, Strategy, and Behavior*, 8th edition. New York: McGraw-Hill.

Magnusson, M. A. and Wydick, B. (2002). How Efficient are Africa's Emerging Stock Markets? *Journal of Development Studies*, 38(4), 141-156.

Madaki, S. A. and Ogunrayewa, M. O. (1999). The New National Housing Policy and Shelter Challenge. *Journal of Environmental Sciences*, 3(1), 74-79.

Malunga, C. and Banda, C. (2004). *Understanding Organizational Sustainability Through African Proverbs: Insights for Leaders and Facilitators*. Washington, DC, US: Pact Publications.

Marlow, A. T. (2012). ISO 14001 Survey: Towards the Future of EMS. http://andrewtmarlow.wordpress.com/2012/06/19/iso-14001-survey-towards-the-future-of-ems/.

Mason, J. P. (1989). The Role of Urbanization in National Development: Bridging the Rural-Urban Divide. *A.I.D. Program Evaluation* (Discussion

Paper No.27). United States: US Agency for International Development (July).

McLuhan, M. (1964). *Understanding Media: The Extensions of Man*. New York: McGraw-Hill.

McSweeney, B. (2009). Incoherent Culture. *European Journal of Cross-Cultural Competence and Management*, 1(1), 22-27.

Mebratu, D. (1998). Sustainability and Sustainable Development: Historical and Conceptual Review. *Environmental Impact Assessment Review*, 18(6), 493-520.

Mercer, J., Kelman, I., Taranis, L., and Suchet-Perason, S. (2010). Framework for Integrating Indigenous and Scientific Knowledge for Disaster Risk Reduction. *Disasters*, 34(1), 214-239.

Moavenzadeh, F. (1976). The Construction Industry in Developing Countries. *Technology Adaptation Program* (Spring 1975). Cambridge, MA: Massachusetts Institute of Technology.

Moore, K. C. (2013). Making Sense by Measuring Arcs: A Teaching Experiment in Angle Measure. *Educational Studies in Mathematics*, 83 (2), 225-245.

Moran, T. H., Graham, E. M. and Blomström, M. (Eds.) (2005). *Does Foreign Direct Investment Promote Development?* Washington DC, US: Institute for International Economics.

Murphy, C. N. and Yates, J. (2009). *International Organization for Standardization (ISO): Global Governance through Voluntary Consensus*. Madison Avenue, NY: Routledge.

Morrow, D. and Rondinelli, D. (2002). Adopting Corporate Environmental Management Systems: Motivations and Results of ISO 14001 and EMAS Certification. *European Management Journal*, 20(2), 159-171.

Mwiti, N. (2012). The Best 72+ African Wise Proverbs and Inspiring Quotes. *Afritorial* (June 22). http://afritorial.com/the-best-72-african-wise-proverbs/.

NBS (National Bureau of Statistics) (2010). *Annual Abstract of Statistics, 2010*. FCT, Abuja, Nigeria: NBS.

Neely, A., Bourne, M. and Kennerley, M. (2000). Performance Measurement System Design: Developing and Testing a Process-Based Approach. *International Journal of Operations and Production Management*, 20(10), 1119-1145.

Neumayer, E. and Perkins, R. (2004). What Explains the Uneven Take-up of ISO 14001 at the Global Level? A Panel-Data Analysis. *Environment and Planning*, 36, 823-839.

NHP (Nigerian Housing Policy) (2011). *The Nigerian National Housing Policy 2011* (draft copy).

Nyam, S. (2008). Urban Housing In Nigeria. In: *The Forum: Habitat for Humanity and Urban Issues* (volume 15, number 1) (pp. 6-7). www. habitat.org/lc/theforum/english/archives/.

Obidegwu, C. F. (1996). Nigeria: Priorities and Prospects for the 1990s. In: A. Y. Yansané (Ed.), *Prospects for Recovery and Sustainable Development in Africa* (pp. 126-161). Westport, Connecticut, London: Greenwood Press.

Ochieng, E. G. and Price, A. D. (2009). Framework for Managing Multicultural Project Teams. *Engineering, Construction and Architectural Management*. 16(6), 527-543.

Odebiyi, S. O. (2010). Sustainable Housing Development in Africa: Nigerian Perspective. *International Business and Management*, 1(1), 23-30.

Odediran, S. J., Opatunji, O. A. and Eghenure, F. O. (2012). Maintenance of Residential Buildings: Users' Practices in Nigeria. *Journal of Emerging Trends in Economics and Management Sciences*, 3(3), 261-265.

Ofori, G. (2000). Challenges of Construction Industries in Developing Countries: Lessons from various Countries. In: *2nd International Conference on Construction in Developing Countries: Challenges Facing the Construction Industry in Developing Countries* (pp. 15-17). Gaborne: Wiley.

Ofori, G. (2006). Revaluing Construction in Developing Countries. A Research Agenda. *Journal of Construction in Developing Countries*, March, 1-16.

Ogunwusi, B. (2013). NHF: Nigerian Workers and an Uninspiring Housing Scheme. *Daily Independent* (September). http://dailyindependentnig.com/ 2013/09/nhf-nigerian-workers-and-an-uninspiring-housing-scheme/.

Ojo, S. O., Adeyemi, A. Y. and Fagbenle, O. I. (2006). The Performance of Traditional Contract Procurement on Housing Projects In Nigeria. *Civil Engineering Dimension*, 8(2), 81-86.

Okonjo-Iweala, N. (2007). Nigeria's Fight for Debt Relief: Tracing the Path. In: *Sustainable Development (Session II: Angelina, Bono and Me: New Vehicles to Engage the Public)* (August 1). Washington, DC: The Brookings Institution.

Okonjo-Iweala, N. and Osafo-Kwaako, P. (2007). Nigeria's Economic Reforms: Progress and Challenges. *Brookings Global Economy and Development* (Working Paper #6). Washington, DC: The Brookings Institution.

Okparaocha, C. (2013). Nigeria's Social Housing Scheme Lacks Defined Policies – NIESV. *Nigerian Tribune* (April 3). www.tribune.com.ng/news 2013/index.php/en/business-package/2012-10-29-11-35-35/property-environment/item/8762-nigeria-s-social-housing-scheme-lacks-defined-policies-niesv.html.

Okpoko, A. I. (1998). Archaeology and the Study of Early Urban Centers in Nigeria. *African Study Monographs*, 19(1), 35-54.

Oladinrin, T. O., Ogunsemi, D. R. and Aje, I. O. (2012). Role of Construction Sector in Economic Growth: Empirical Evidence from Nigeria. *African Journals Online (Journal of the Environment)*, 7(1), 50-60.

Olayiwola, L. M., Adeleye, O. and Ogunshakin, L. (2005). Public Housing Delivery In Nigeria: Problems and Challenges. In *XXXIII World Congress on Housing: Transforming Housing Environments though the Design*. Pretoria, South Africa (September 27-30).

Oluwakiyesi, T. (2011). *Construction Industry Report: A Haven of Opportunities*. Lagos, Nigeria: Vevita Capital Management Limited.

Ong, S. E. (2000). Housing Affordability and Upward Mobility from Public to Private Housing in Singapore. *International Real Estate Review*, 3(1), 49-64.

Orji, U. (2013). Why Nigeria May Not Meet the Millennium Development Goals on Housing. *Peoples Daily* (March 12). http://peoplesdailyng.com/why-nigeria-may-not-meet-the-millennium-development-goals-on-housing-2/.

Orts, E. and Spigonardo, J. (2012). *Greening the Supply Chain: Best Practices and Future Trends*. US: The Initiative for Global Environmental Leadership and Knowledge @ Wharton.

Osagie, C. (2011). Nigeria-China Trade Hits $7.76bn. *This Day Live* (April 5). www.thisdaylive.com/articles/nigeria-china-trade-hits-7-76bn/89066/.

Osborne, R. (2005). Urban Sprawl: What is Urbanization and Why does it Matter? In: *Proceedings of the British Academy*, 126 (pp. 1-16). Oxford University Press.

Osagie, C. (2012). SON Pursues 300% Rise in SME Quality Assurance. *This Day Live* (May 1). www.thisdaylive.com/articles/son-pursues-300-rise-in-sme-quality-assurance/114860/.

Osasona, C. O. (2007). From Traditional Residential Architecture to the Vernacular: The Nigerian Experience. www.mudonline.org/aat/2007_documents/AAT_Osasona_web-based%20publication_060928.pdf.

Oyebade, W. (2013). FG, Lagos Unfold Incentives for Mass Housing Investors. *The Guardian* (November 17). www.ngrguardiannews.com/

index.php/news/national-news/138778-fg-lagos-unfold-incentives-for-mass-housing-investors.

Oyejide, T. A. (1991). Adjustment with Growth: Nigerian Experience with Structural Adjustment Policy Reform. *Journal of International Development*, 3(4), 485-498.

Pearce, D. W. and Atkinson, G. D. (1993). Capital Theory and the Measurement of Sustainable Development: An Indicator of "Weak" Sustainability. *Ecological Economics*, 8(2), 103-108.

Pepple, A. I. (2012). Federal Ministry of Lands, Housing and Urban Development. *A presentation at the Ministerial Platform*. Abuja, Nigeria: Federal Ministry of Information (May 15). http://fmi.gov.ng/wp-content/uploads/2011/10/Federal-Ministry-of-Lands-Housing-and-Urban-Development.pdf.

Phang, S. Y. and Wong, W. K. (1997). Government Policies and Private Housing Prices in Singapore. *Urban Studies*, 34(11), 1819-1829.

PM News (2013). Chinese Firm to Build 2000 Houses in Ogun State. *PM News* (September 25). http://pmnewsnigeria.com/2013/09/25/chinese-firm-to-build-2000-houses-in-ogun-state/.

Posthuma, R. A. (2009). National Culture and Union Membership: A Cultural-Cognitive Perspective. *Industrial Relations*, 64(3), 507-529.

Postman, N. (1992). Technopoly: The Surrender of Culture to Technology. New York, US: Vintage Books.

Pratihar, D. K. (2008). *Soft Computing*. Oxford, United Kingdom: Alpha Science International Ltd.

Premium Times (2013). Nigeria to Conduct Housing Census to Ascertain Housing Deficit – Minister. *Premium Times* (July 22). http://premium timesng.com/news/141471-nigeria-to-conduct-housing-census-to-ascertain-housing-deficit-minister.html.

Quaddus, M. A. and Siddique, M. A. B. (Eds.) (2013). *Handbook of Sustainable Development Planning: Studies in Modeling and Decision Support* (2nd edition). Cheltenham, UK: Edward Elgar Publishing Limited.

Queiros, C. (2008). The Complexities of Delivering Urban Housing Projects. In: *The Forum: Habitat for Humanity and Urban Issues* (volume 15, number 1) (pp. 4-5). www.habitat.org/lc/theforum/english/urban/Urban_housing_Nigeria.aspx.

Radelet, S. (2010). *Emerging Africa: How 17 Countries are Leading the Way*. Washington, DC: Center for Global Development.

Raines, S. S. (2002). Implementing ISO 14001 – An International Survey Assessing the Benefits of Certification. *Corporate Environmental Strategy*, 9(4), 418-426.

Rau, H. and Fahy, F. (2013). Sustainability Research in the Social Sciences – Concepts, Methodologies and the Challenge of Interdsciplinarity. In: F. Fahy and and H. Rau (Eds.), *Methods of Sustainability Research in the Social Sciences* (pp. 3-24). London: SAGE.

Reckford, J. (2006). Defining Sustainability. In: *The Forum: Environmental Impact on Housing and Construction Innovations* (volume 13, number 2) (p. 4). www.habitat.org/lc/theforum/english/archives/.

Reina, P. and Tulacs, G. J. (2013). *The Top 250 International Contractors*. New York: Engineering News-Record.

Rotberg, R. I. (2013). *Africa Emerges: Consummate Challenges, Abundant Opportunities*. Cambridge, UK: Polity Press.

Rowley, J. (2002). Using Case Studies in Research. *Management Research News*, 25(1), 16-27.

Scandura, T. A. and Williams, E. A. (2000). Research Methodology in Management: Current Practices, Trends and Implications for Future Research. *The Academy of Management Journal*, 43(6), 1248-1264.

Sennara, M. and Hartman, F. (2002). Managing Cultural Risks on International Projects. In: *Proceedings of the Project Management Institute Annual Seminars and Symposium*. San Antonio, Texas, US (October 3-10).

Sharifi, S. and Behnoud, E. (2013). The Relation Between Sustainable Development and Sustainable Architecture. *Life Science Journal*, 10(2), 114-117.

Srinivasan, S. (2009). The Emerging Markets Century: How a New Breed of World-Class Companies is Overtaking the World. *Great Lakes Herald*, 3 (2), 64-66.

Ssegawa, J. K. (1999). Construction Industry's Contribution towards Botswana's Vision 2016. In: W. Hughes (Ed.), *15th Annual ARCOM Conference* (pp. 743-52), September 15-17, Association of Researchers in Construction Management, Liverpool John Moores University.

Storti, C. (1998). *Figuring Foreigners Out: A Practical Guide*. US: Intercultural Press Inc.

Take, O. (1984). Abuja, the New Capital of Nigeria, and the Urban Design of its Central Area. In: M. B. Ševčenko (Ed.), *Designing in Islamic Cultures 4: Continuity and Change: Design Strategies for Large-Scale Urban*

Development (pp. 49-55). Cambridge, MA: The Aga Khan Program for Islamic Architecture.

Taylor, I. (2006). Unpacking China's Resource Diplomacy in Africa. *Center on China's Transnational Relations* (Working Paper No. 19). Hong Kong: The Hong Kong University of Science and Technology.

Tellis, W. (1997). *Application of a Cse Study Methodology. The Qualitative Report*, 3(3). www.nova.edu/ssss/QR/QR3-3/tellis2.html.

Tessema, F., Taipale, K. and Bethge, J. (2009). *Sustainable Building and Construction in Africa*. Germany: Federal Ministry for the Environment, Nature Conservation and Nuclear Safety.

The Economist (2013). Why is South Africa included in the BRICS? *The Economist* (March 29). www.economist.com/blogs/economist-explains/2013/03/economist-explains-why-south-africa-brics.

Tisdale, H. (1947). The Process of Urbanization. *Social Forces*, 20(3), 311-316.

Top Asia News (2012). Nigerian Finance Minister to Meet Chinese Bank Officials Over 3bn-dollar Loans. *Top Asia News* (February 23). www.topasianews.com/news/Nigerian-finance-minister-to-meet-Chinese-bank-officials-over-3bn-dollar-loans.html.

Tuna, A. (2013). A Conceptual Analysis of the Knowledge of Prospective Mathematics Teachers about Degree and Radian. *World Journal of Education*, 3(4), 1-9.

Tylor, E. B. (1871). *Primitive Culture: Researches Into the Development of Mythology, Philosophy, Religion, Art, and Custom*. In: Two Volumes (Vol. II). Albemarle Street, London: John Murray.

Udo, B. (2013). Nigeria Targets 200,000 Low-Cost Housing Units in Five Years. *Premium Times* (May 29). http://premiumtimesng.com/business/136886-nigeria-targets-200000-low-cost-housing-units-in-five-years.html.

Ukoha, O. M. and Beamish, J. O. (1996). Predictors of Housing Satisfaction in Abuja, Nigeria. *Housing and Society*, 23(3), 26-46.

UN-HABITAT (2003). *The Challenge of Slums: Global Report on Human Settlements 2003*. UK and US: Earthscan Publications Ltd. www.unhabitat.org/content.asp?typeid=19andcatid=555andcid=5373.

UNDP (United Nations Development Program) (2011). *Towards Human Resilience: Sustaining MDG Progress in an Age of Economic Uncertainty*. New York, US: UNDP Bureau for Development Policy.

UNFPA (United Nations Population Fund) (2011). *State of World Population 2011: People and Possibilities in a World of 7 Billion*. New York, US: UNFPA.

UN System Task Team (2012). Culture: A Driver and an Enabler of Sustainable Development. *Thematic Think Piece* (UN System Task Team on the Post-2015 UN Development Agenda, May). Paris, France: UNESCO.

Uroko, C. (2013). Summit on Housing Deficit to Consider Dry Construction Option. *Business Day* (January 29). http://businessdayonline.com/NG/index.php/homes-a-property/50784-summit-on-housing-deficit-to-consider-dry-construction-option.

Van Binsbergen, W. M. J. (2009). Expressions of Traditional Wisdom from Africa and Beyond: An Exploration in Intercultural Epistemology. Brussels, Belgium: Academie Royale Des Sciences D'outre-Mer.

Waziri, B. S. and Vanduhe, B. A. (2013). Evaluation of Factors Affecting Residential Building Maintenance in Nigeria: Users' Perspective. *Civil and Environmental Research*, 3(8), 19-24.

Webster, J. and Watson, R. T. (2002). Analyzing the Past to Prepare for the Future: Writing a Literature Review. *MIS Quarterly*, 26(2), xiii-xxiii.

Whiten, A. and van Shaik, C. P. (2007). The Evolution of Animal 'Cultures' and Social Intelligence. In: N. Emery, N. Clayton and C. Frith (Eds.), *Social Intelligence: From Brain to Culture* (pp. 189-216). New York: Oxford University Press.

Wibowo, M. A. (2009). The Contribution of the Construction Industry to the Economy of Indonesia: A Systemic Approach. Department of Civil Engineering, Diponegoro University, Indonesia.

Windapo, A. O. and Rotimi, J. O. (2012). Contemporary Issues in Building Collapse and Its Implications for Sustainable Development. *Buildings*, 2, 283-299.

Wolgin, J. M. (1999). The Foundations of Sustainable Development in Africa: Ten Trends and One Conundrum. In: V. U. James (Ed.), *Sustainable Development in Africa: Prospects for the 21st Century* (pp. 9-22). San Francisco: International Scholars Publications.

Woo, S. B. (2013). Over 200,000 Public, Private Homes to be Complete by 2016. *Today Online* (October 26). www.todayonline.com/singapore/over-200000-public-private-homes-be-completed-2016.

World Bank (2013). What are Emerging Markets? https://datahelpdesk.worldbank.org/knowledgebase/articles/114958-what-are-emerging-markets-.

Zadek, S., Chen, X., Jia, T., Zhou, Y., Yu, K., Forstater, M., and Morgan, G. (2009). Responsible Business in Africa: Chinese Business Leaders' Perspectives on Performance and Enhancement Opportunities. *Corporate*

Social Responsibility Initiative (Working Paper No. 54). Cambridge, MA, US: Harvard Kennedy School of Government.

Zeng, S. X., Tian, P. and Shi, J. (2005). Implementing Integration of ISO 9001 and ISO 14001 for Construction. *Managerial Auditing Journal*, 20(4), 394-407.

In: Urbanization
Editor: Harold J. Kauffman ISBN: 978-1-63117-063-8

Chapter 5

URBANIZATION AND CARDIOMYOPATHY VULNERABILITY IN AN AFRICAN MALE COHORT: THE SYMPATHETIC ACTIVITY AND AMBULATORY BLOOD PRESSURE IN AFRICANS STUDY

R. P. Van Staden, L. Malan* and J. M. Van Rooyen

Hypertension in Africa Research Team (HART), North-West University, Potchefstroom Campus, Potchefstroom, North West Province, South Africa

ABSTRACT

Objective: Vulnerability for hypertension in Africans increases when they are exposed to psychosocial stress or urbanization. Inevitable cultural and lifestyle changes may impact on cardiomyopathy risk markers. Therefore, we aimed to assess the relationship between blood pressure, silent ischemia, glycated hemoglobin (HbA1c), QTc dispersion

* Corresponding author: Prof Leoné Malan (RN, PHD). Address: Hypertension in Africa Research Team (HART), North-West University, Potchefstroom Campus, Private Bag X6001, Potchefstroom, 2520, North-West, South Africa. Tel: (+27) 18 299 2438. Fax: (+27) 18 299 1053. E-mail: leone.malan@nwu.ac.za.

and ECG Cornell Product left ventricular hypertrophy (LVH) in a cohort of African men.

Methods: We included 100 Caucasian and 82 African men, aged 25-60 years, from the Sympathetic activity and Ambulatory Blood Pressure in Africans (SABPA) Study. HbA1c as well as ambulatory blood pressure (BP) and -silent ischemia values were evaluated, compared and correlated. QTc dispersion and Cornell product data were obtained via 12-lead ECG.

Results: African men revealed ($p \leq 0.05$) pre-diabetic status with a mean HbA1c (> 6%). In addition, they revealed increased ambulatory BP, silent ischemia, QTc and LVH values compared to the Caucasian men. In forward stepwise regression analyses [Adjusted R^2 for model, 0.33], ambulatory systolic BP (0.53, 95% CI 0.12 to 0.94) and silent ischemia (0.26, 95% CI 0.06 to 0.46) were associated with LVH in African men.

Conclusion: Systolic BP and silent ischemia were associated with LVH in a cohort of African men. Increased cardiovascular disease risk markers may reveal progressing cardiomyopathy.

INTRODUCTION

Africans and their vulnerability to hypertension has been the topic for many research studies. [1] Although hypertensive ethnic treatment-regimes have improved, it does not seem to have the impact it should to effectively control blood pressure in the African population [2] This may have an effect on the myocardium even though ischemic heart disease is still relatively uncommon. Currently urban Africans share the same cardiovascular risk factors with Caucasians such as substance abuse, elevated cholesterol and physically inactivity. [2] Pickering et al. [3] discussed a few viewpoints on racial differences in disease where hypertension was the main theme and they favored environmental factors [1,2] as the main cause for the differences.

Hypertension shares the same risk factors as those for left ventricular failure. It can start with reversible factors such as increased alcohol consumption and abnormal glucose levels. [4] Abnormal levels of glucose and HbA1c are associated with increased cardiovascular disease risk. [5] Kubiak et al. [6] ventured further and established some support for the role of sympathetic activation and the QTc dispersion-glucose link. This link was explored by Ishikawa et al. [7] where their study proposed that Cornell product left ventricular hypertrophy and QTc dispersion is related to glucose abnormality and has a predictive value for stroke and cardiovascular disease. ECG markers, such as the Cornell product, can therefore be used to identify possible pathology early and is an easy, relatively non-expensive non-invasive procedure.

Malan et al. [8] supported the findings of Pickering et al. [3] where psychosocial stress was shown to have an impact on Africans' cardiovascular risk. The hypertensive prone state of the Africans exposes them to other pathologies such as diabetes, stroke, hypertension and sudden cardiac death. [9] Our aim was, therefore, to assess the association between HbA1c, QTc dispersion and left ventricular hypertrophy in African and Caucasian men residing in an urban area.

MATERIALS AND METHODS

Study Design, Population and Protocol

The Sympathetic activity and Ambulatory Blood Pressure in Africans Study was a target population comparative study conducted in 2008 – 2009 avoiding seasonal changes. The target population study included 409 urban African and Caucasian male and female teachers, aged 25-60 years. All participants were from the Dr Kenneth Kaunda Education district, in the North-West Province of South Africa and thus shared similar socio-economic status. Men are the gender at greater risk for cardiovascular disease [10] and therefore in our sub-study we only included the data from African (N=101) and Caucasian (N=100) men. Eligibility to participate was screened by a registered nurse. Exclusion criteria were tympanum temperature > 37.5°C, vaccinated/blood donors 3 months prior and medication (alpha-, beta blockers and psychotropic substance abuse). We also excluded men should they have a HIV positive status (N=13) or been clinically diagnosed as diabetics (N=8). Our subsample of the SABPA study therefore comprised 82 African and 100 Caucasian men. Written informed consent was obtained from all participants after the procedures were explained to them in detail in their home language and they were given the opportunity to ask questions.

From 0700 – 0800 every day of the working week, four subjects were fitted with the Cardiotens® (ambulatory blood pressure and -ECG apparatus) and a physical activity meter (Actical®, Mini Mitter, Montreal, Quebec) and the software was activated. Thereafter, the teachers resumed their normal daily activities until ~ 1600. At this point in time they were transported to the North-West University where the Metabolic Research Unit is situated. The unit is well equipped with 10 bedrooms and various facilities to facilitate a comfortable overnight. On arrival, the participants were welcomed and received their respective rooms. They were introduced to the experimental setup to lessen anticipation stress and the participants received pre-counseling for HIV/AIDS by a trained nurse. They were advised to go to bed at ± 2200 and to fast overnight. At 0600 the following morning, after the last ambulatory blood pressure and -ECG were recorded, the apparatuses were removed and anthropometric measurements taken. Participants were in the semi-Fowlers position for a resting 12-lead ECG measurement obtained by a registered nurse and physician within two identical BP stations. Blood sampling followed. Ambient temperature, noise and illumination were carefully controlled.

Feedback and post-counseling were done in the privacy of their own rooms regarding their anthropometric, blood glucose measurements and HIV status. Ambulatory blood pressure and -ECG reports were sent to them after one week and the participants were notified timely for possible referral to their doctors where necessary.

Lifestyle Factors and Anthropometric Measurements

Physical activity was measured using the Actical® omnidirectional accelerometer which was fitted on the hip of the participant. Set to 15 second intervals, this device measured movement of the participant whilst they engaged in normal daily activities. The data was then converted into 1-second intervals for analysis. γ-GT levels were measured as marker for alcohol abuse, and for each participant's smoking habits, cotinine was measured. All anthropometric measurements were obtained by level 2 accredited anthropometrists using calibrated instruments (Precision health scale, A & D Company, Tokyo, Japan; Invicta Stadiometer IP 1465, Invicta, London, UK), with subjects in minimal clothing and without shoes. Intra- and inter-observer variability was less than 10%. Measurements were performed with standardized methods and in triplicate to ensure accuracy. Body mass index (BMI) was calculated by dividing the participant's weight (kg) by their length (m^2).

Blood Pressure

A trained registered nurse and medical doctor obtained all cardiovascular measurements. The ambulatory blood pressure measurements and 12-lead electrocardiogram measurements were done with the Cardiotens® CE120 (Meditech, Hungary), validated by the British Hypertension Society. The apparatus and leads were fitted each morning ~ 0700 – 0800, applying suitable cuff sizes on the non-dominant arm. The Cardiotens measured blood pressure at 30-min intervals during the day (0800 – 2200) and 60-min intervals at night (2200 – 0600 hours). Successful mean inflation rates were 82.7 % (±3.8%) in African and 94.6% (±3.7%) in Caucasian men for the 22 – 23 hour ambulatory blood pressure period. Participants were asked to continue with their normal daily activities and record any abnormalities such as visual disturbances, headache, nausea, fainting, palpitations, physical activity and emotional stress on their ambulatory diary cards. The data were analyzed using the Cardio-Visions 1.15 Personal Edition software (Meditech®).

The Ischemic Profile, Left Ventricular Hypertrophy (LVH) and QTc Dispersion

The *silent ischemic profile*: was assessed by two-channel ambulatory ECG recordings (Cardiotens® CE120) according to a preset program for 20s at 5-min intervals. Silent ischemic events were recorded according to the 1-1-1 rule which entails a horizontal or descending ST-segment depression by one mm; duration of the ST-segment episode of at least one minute and a one minute interval from previous episodes. A resting 12-lead ECG (strip lead II) obtained ST segment deviations during rest (NORAV Medical Ltd PC 1200, software version 5.030, Israel). ST depression was defined when a ST segment deviation of 1 mm was evident in the limb or chest recordings V1 to V6 in six cardiac cycles. *ECG LVH*: A 12-lead ECG determined LVH using strip leads RaVL + SV3 in the calculation of a gender-specific formula, the Cornell product: sum of leads (RaVL+SV3)*QRS duration, where a value of > 244 mV.ms is considered as left ventricular hypertrophy.

The *QTc dispersion*: was calculated by the equation: QTc max-QTc min = ECG Resting QT/QTC (dispersion) in milliseconds. (NORAV Medical Ltd PC 1200, Software version 5.030, Israel).

Blood Sampling and Biochemical Analysis

Resting fasting blood samples were obtained by a registered nurse from the brachial vein branches using a winged infusion set with syringes for preparation of serum, EDTA plasma and NaF samples. All fasting blood samples were handled according to the appropriate methods. For HIV/AIDS, screening tests included an antibody test: a A rapid card test, First Response (PMC Medical, India) was used for testing and if they tested positive, the result was repeated with the Pareeshak (BHAT Bio-tech India) card test for confirmation.

NaF glucose was determined with the timed-end point method with Unicel DXL 800 – Beckman & Coulter, Germany. The liver enzyme, serum γ-glutamyl transferase (U/l) and lipogram profiles, were assessed with an enzyme rate method (Unicel DXC 800 - Beckman and Coulter, Germany). Cotinine levels were determined by homogenous immunoassay with the Roche Modular System (Roche, Basil, Switzerland). EDTA whole blood turbidometric inhibition immunoassay analyses determined HbA1c (glycated hemoglobin) which reflects an average capillary glucose equivalent for the preceding 2-3 months (Integra 400 Roche, Switserland).

Data Management and Statistical Analyses

Statistical analyses were done with Statistica software v9.0. Shapiro Wilks analysis determined normal distributions. Chi-square tests were used to determined proportions. Univariate analysis determined differences between groups and confounders. Subsequently, one way ANCOVA's followed adjusting for covariates (physical activity and γ-glutamyl transferase). Multiple regression analysis was performed for all cardiovascular risks factors in each of the ethnic groups. Stepwise forward regression analysis was performed with LVH as dependent variable. Independent variables included HbA1c, resting ECG QTc dispersion, ambulatory SBP, silent ischemia, physical activity and γ-glutamyl transferase. P-values of ≤ 0.05 were considered as statistically significant.

RESULTS

In Table 1, it is clear that Africans tend to be physically more inactive as well as consuming more alcohol as depicted by GGT levels. In Table 1 and in Figure 1, the African men revealed a more vulnerable cardiometabolic risk profile compared to their Caucasian counterparts. An HbA1c above 6.5% is a newly recommended criterion for diagnosing diabetes. Despite the protective mean lower cholesterol levels, African men ($p \leq 0.05$) demonstrated a mean pre-diabetic state HbA1c (6.2%). In comparison to the Caucasian men, they demonstrated increased ambulatory BP, heart rate, LVH, QTc dispersion values and more silent ischemic events.

Table 1. Participant characteristics (mean ± SD) by ethnic status

Parameters	African men (N = 82)	Caucasian men (N = 100)	p-Value
Age, y	43.2 ± 8.10	45.0 ± 11.14	0.190
Lifestyle factors			
BMI (kg/m^2)	27.6 ± 5.83	29.0 ± 5.20	0.060
Cotinin (ng/mℓ)	35.5 ± 65.01	30.9 ± 96.72	0.690
Physical activity (kcal/24h)	2714.9 ± 800.12	3674.4 ± 2059.23	< 0.001
γ-glutamyl transferase (U/l)	84.8 ± 91.71	34.7 ± 29.52	< 0.001
Cardiovascular variables			
SBP (mmHg)	138 ± 16	128 ± 10	< 0.001
DBP (mmHg)	88 ± 11	80 ± 70	< 0.001
Heart rate (beats per minute)	78.8 ± 11.3	71.8 ± 11.1	< 0.001
HbA1c (%)	6.2 ± 1.23	5.7 ± 0.48	< 0.001
Cholesterol	4.74 ±1.17	5.58 ± 1.20	< 0.001
HdL cholesterol	1.05 ± 0.37	1.00 ± 0.27	0.225
Triglycerides	1.81 ± 1.59	1.51 ± 0.87	0.118
ST depression N (%)	38 (56.7)	29 (43.28)	0.003
Hypertensive N (%)	63 (76.8)	69 (69.0)	0.160
Medications			
Hypercholesterolemia N (%)	1 (14.3)	6 (85.7)	0.049
Hypertension N (%)	14 (17.1)	9 (9.0)	0.027

Where: N, number of participants; BMI, body mass index; γ-glutamyl transferase, cGGT; HbA1c, glycated hemoglobin. **Significant values depicted as $p \leq 0.05$.**

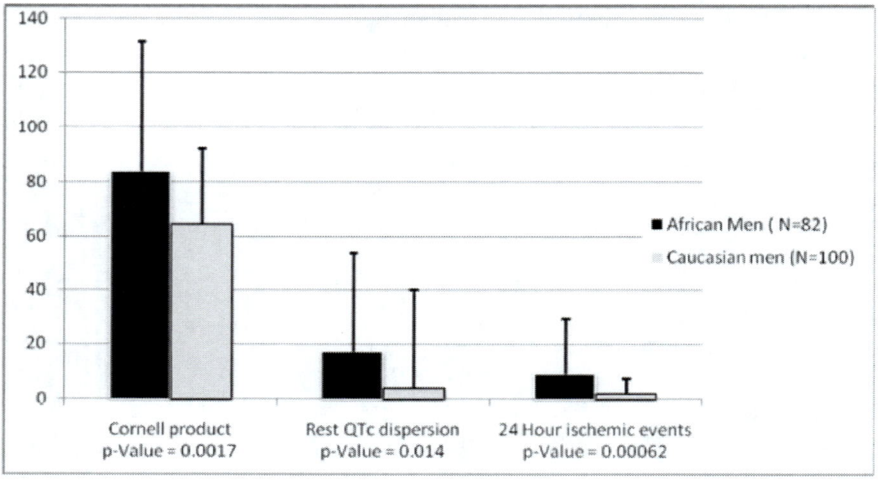

Figure 1. The cardiovascular profile (mean ± SE) of African and Caucasian men independent of physical activity and log γ-glutamyl transferase.

In Table 2, significant associations only existed between LVH, ambulatory SBP and silent ischemia in the African men but not in Caucasian men. Therefore we computed forward stepwise regression analyses in African men only (Table 3). Positive associations were demonstrated between LVH, ambulatory SBP [0.53 (95% CI 0.12 to 0.94)] and silent ischemia [0.26 (95% CI 0.06 to 0.46)] [Adjusted R^2 for model, 0.33].

Table 2. Partial correlations between cardiovascular measures and ECG left ventricular hypertrophy (LVH) as well as QTc dispersion in African men

	African men (N=82)			
	ECG LVH		QTc	
Parameter	r - Values	p- Values	r - Values	p-Values
Ambulatory SBP	0.47	<0.001	0.20	0.080
Ambulatory silent ischemia	0.39	0.002	0.17	0.130
Resting QTc dispersion	0.10	0.402	-	-
HbA1c	0.02	0.854	0.24	0.030

Where: ECG LVH, utilizing the Cornell product (RaVL+SV3)*QRS duration. Data were adjusted for physical activity and log γ-glutamyl transferase.

Table 3. Forward stepwise regression analyses results of the relationship between ECG left ventricular hypertrophy and cardiovascular variables in African men

	ECG Left ventricular hypertrophy	
Parameters	*African men (N=82)*	P- values
Adjusted R^2	0.33	
	β (95% CI)	
Ambulatory SBP	0.53 (0.12, 0.94)	0.010
Ambulatory silent ischemia	0.26 (0.06, 0.46)	0.001

Covariates considered for the model included HbA1c, QTc dispersion, physical activity and log γ-glutamyl transferase.

Discussion

The aim of the study was to assess HbA1c, increased QTc dispersion and its association with left ventricular hypertrophy (LVH). Analysis of data though revealed no significant associations between all mentioned markers i.e., QTc dispersion, elevated HbA1c and left ventricular hypertrophy. Findings suggested that, although the HbA1c was not associated with left ventricular hypertrophy, strong associations were revealed between left ventricular hypertrophy and some of the hallmark risk factors for stroke and hypertension in the African men. [10] Risk factors included elevated blood pressure and abnormal QTc dispersion [11] in the presence of silent ischemia. On a related note, control of target blood pressure levels in the African men on anti-hypertensive drugs is poor compared to the Caucasian men. This may impact on the myocardium even though ischemic heart disease in Africans is still relatively uncommon. [1] Our data therefore could support the pathological pathway in the progression of cardiomyopathy. [12] Other alarming factors present in the African men included elevated HbA1c, physical inactivity and alcohol abuse, which also have been associated with cardiovascular disease. [13]

Ambulatory systolic blood pressure was the strongest predictor for LVH in our study. Epidemiologically it was shown that the risk for coronary artery disease (CAD) and stroke death increase curvilinearly [12] with pronounced increases in systolic blood pressure (SBP). Although treatment for systolic hypertension has been well supported [14] it deserves more attention in the African population. Conversely, SBP impacts on CAD when abnormally elevated, and in turn, is known to cause ischemia in the myocardial tissue. [15] Ischemia implies a lack in oxygen delivery to the myocardial tissue and this will cause a decrease in ATP production. [16] The energy deficiency will lead to adaptation of the myocytes that will manifest as hypertrophied tissue. [16, 20] The presence of ischemia was supported by the numerous silent ischemic events found in the African men. Their higher systolic blood pressure could contribute to hypertension and progression of arterial stiffness. [17]

Hinderliter et al. [18] contradicted Middlemost [19] in revealing that diastolic BP or vascular reactivity correlated better with LVH in Africans than arterial stiffness. Opie et al. [20] however, supported the diastolic dysfunction effect of Africans with LVH. He suggested that the loss of distensability implies a stiffer ventricle, impaired relaxation with a decreased early diastolic filling in LVH in Africans. Subsequently, the delivery of blood to the coronary arterial circulation during diastole will be compromised. [18-20] Decreases in coronary perfusion will augment the already existing ischemic situation. [20]

Recently other studies described the environmental impact on Africans residing in an urban area. [1, 21] The influences of chronic psychosocial and acute stressors yielded an increase in sympathetic activity. [8] Increased sympathetic activity not only impacts on sodium retention but also increases heart rate, cardiac output and systolic blood pressure. [1] Increased sympathetic nervous system activity could therefore be a key factor in hemodynamic abnormalities namely increased cardiac output and stiffening of the aorta, especially in systolic hypertension. [20]

Where cardiovascular disease is evident, the accurate electrocardiographic diagnosis of LVH is important to detect hypertrophy in association with underlying cardiovascular diseases (CVD). [18, 19] The ECG plays an important role in the assessment of the initial appearance of various CVD and to monitor the progression or recession of hypertrophy during therapeutic intervention. [20] Our data only indicated a direct association between QTc dispersion and HbA1c in the African men. QTc dispersion is the maximum QTc – minimum QTc interval. [22] The QT interval, which is the traditional measurement for assessing the duration of ventricular de- and repolarization, seems to be affected in pathologies such as hypertrophic cardiomyopathy during the depolarizing period. [22, 23] A prolonged QT interval has been suggested to increase mortality when assessed with an electrocardiogram. [23] Increased QTc dispersion shows a greater variability in myocardial repolarization. [23, 24] The QT segment represents the QRS complex, the ST segment and the T wave and during ischemia all of these variables will change. [24] Ischemic ST elevation and hyper-acute T wave changes are of the earliest signs of ischemia and infarction. [24] The QRS complex changes in acute and chronic ischemia although we found no significant associations between these markers. ECG manifestations will be the possible presence of the pathological Q wave with loss of R wave and broad complex or taller R waves and deep S waves. [12] This may be due to LVH where an increase occurs in electrical forces which are directed to the left. The vectors of these forces are very prominent in V1 and V6 whilst in LVH the R wave in V6 and the S wave in V1 are greatly accentuated, whereas right ventricular failure increases both waves in both leads.

Kawasaki et al. [25] demonstrated that changes in the repolarization period did not correlate strongly with increased QTc dispersion but rather with the left ventricular portion of maximum hypertrophy. They also concluded inconsistently that there is no correlation between QTc dispersion and established markers of risk for cardiovascular disease. However, their study population included multi-ethnic groups. In our study, QTc dispersion did not predict LVH and our findings are supported by Chapman et al. [26] where a prolonged QTc was observed in Africans compared to Caucasians.

The high levels of HbA1c in the Africans were associated with QTc dispersion but not with LVH. If we consider the increased HbA1c, SBP, QTc dispersion, low physical activity and possible alcohol abuse in the African men it could strengthen a proposed pathway for progression of LVH. [27] Hyperglycemia has been strongly associated with adverse outcomes in patients hospitalized with ventricular hypertrophy and acute coronary syndrome.]28] In a recent study by Selvin et al. [29], glycated hemoglobin was found to contribute to increased risk for diabetes, cardiovascular disease, stroke and mortality. [30] If there is an additional lack of physical activity and alcohol abuse, glucose control is difficult, apart from the positive effects of exercise on systems such as the adipocyte-endocrine system. [31-32] What could be confounding the HbA1c findings, are markers such as anemia and erythrocyte abnormalities, which are not known in our sub study. Apart from the SBP and QTc dispersion associated with LVH, our data revealed augmented hypertension and stroke risk in the African male cohort. [32-33] Hypertension and stroke not only contributes greatly to the burden of disease, it also places financial burden on Africans and their families. Stroke in African American population in the U.S. is 1.5 times more prevalent than in the Caucasian population. [10] Hypertension plays a pivotal role when associated with other pathologies in stroke cases. If treated correctly, decreases in the incidence of stroke, heart diseases such as LVH and kidney failure could prevail.

As left ventricular hypertrophy is associated with ventricular arrhythmias and sudden death as a risk, [34] early detection is of the utmost importance and cannot be accentuated enough. A limitation with QTc dispersion is that when QTc minimum is also increased, the QTc dispersion may be insignificant but the damage to the myocardium stay real. The authors also favor the increased maximum QTc, QTc interval or QT/RR ratio as cardiovascular risk markers.

Another limitation is that our study design only included cross sectional data and no causality can be inferred. We recommend that the study be expanded in a follow-up study to assess the possible vulnerability of Africans to cardiovascular disease and progressing cardiopyopathy.

Once the problem of ethnic/racial differences in the pathology of LVH is better described in aetiological terms of environmental influences, it could benefit the patient in both the realms of research and practice.

CONCLUSION

The African men compared to the Caucasian men, both residing in an urban area, are at greater risk for undiagnosed and progression of target organ damage. Systolic BP and silent ischemia were associated with LVH in an African male cohort revealing progressing cardiomyopathy.

Why the risk is not similar for both ethnic groups, is unclear. Societal-Cultural adaptation may be a factor but prospective studies are urgently needed to clarify the matter.

ACKNOWLEDGMENTS

North-West University, National Research Foundation, Medical Research Council, North-West Department of Education, ROCHE Diagnostics, South Africa; Metabolic Syndrome Institute, France. The funding organizations played no role in the design and conduct of the study; collection, management, analysis and interpretation of the data; preparation, review, or approval of the manuscript.

REFERENCES

[1] Seedat YK. Perspectives on research in hypertension. *Cardiovasc J. Afr.* 2009; 20: 39-42.

[2] Cooper R, Rotimi C. Hypertension in blacks. *Am. J. Hypertens.* 1997; 10:804-812.

[3] Pickering T G. Why is hypertension more common in African Americans? *J. Clin. Hypertens.* 2001; 3:50-52.

[4] Peter R, Okoseime O E, Rees A, Owns D R. Postprandial glucose - A potential therapeutic target to reduce cardiovascular mortality. *Current Vasc. Pharm.* 2009; 7:68-74.

[5] Chapman M J, Sposito A C. Hypertension and dyslipidaemia in obesity and insulin resistance: pathophysiology, impact on atherosclerotic disease and pharmacotherapy. *Pharm. & Therap.* 2008; 117: 354-373.

[6] Kubiak T, Wittig A, Koll C, Mraz B, Gustav J, Herrmann U, Weber H, Kerner W. Continuous glucose monitoring reveals associations of glucose levels with QT interval length. *Diab. Tech. & Therap.* 2010; 12:283-286.

[7] Ishikawa J, Ishikawa S, Kabutoya T, Gotoh T, Kayaba K, Schwartz J E, Pickering T G, Shimada K, Kario K. Cornell product left ventricular hypertrophy in electrocardiogram and the risk of stroke in a general population. *Hypertension* 2009; 53:28-34.

[8] Malan L, Malan N T, Du Plessis A, Wissing M P, Potgieter J C, Seedat Y K. The cost of coping: a cardio-neuro-metabolic risk for black South Africans? *Cardiovasc. J. Afr.* 2010; 21:183-185.

[9] Primatesta PBL, Poulter NR. Blood pressure levels and hypertension status among ethnic groups in England. *J. Hum. Hypertens.* 2000; 14:143-148.

[10] Moulton S A. Hypertension in African Americans and its Related Chronic Diseases. *J. Cult. Diversity* 2009;16:65-170.

[11] Padmanabhan S, Silvet H, Amin J, Pai R G. Prognostic value of QT interval and QT dispersion in patients with left ventricular systolic dysfunction: results from a cohort of 2265 patients with an ejection fraction of < or =40%. *Am. Heart J.* 2003;145:132-138.

[12] Libby P, Bonow R O, Mann D K, Zipes D P: Braunwald's Heart Disase: *A textbook of Cardiovascular Medicine Philadelphia,* Saunders Elsevier. 2008.

[13] Suliman S, Seedat S, Williams D R, Stein D J. *Predictors of transitions across stages of alcohol use and alcohol-use disorders in South Africa.* 2010;71:695-703.

[14] Volpe M. Treatment of systolic hypertension: spotlight on recent studies with angiotensin II antagonists. *J. Hum. Hypertens.* 2005; 19:93-102.

[15] Cameron J S, Bassett A L, Gaide M S, Lodge N J, Wong S S, Kozlovskis P L, Myerburg R J. Cellular electrophysiology of coronary artery ligation in chronic pressure overload. *Int. J. Cardiol.* 1987; 14:155-168.

[16] Opie L H. *HEART Physiology: From Cell to Circulation*. Vol. 4 Philadelphia,Lippincott-Raven Publishers,2004.

[17] Triantafyllidi H, Tzortzis S, Lekakis J, Ikonomidis I, Arvaniti C, Trivilou P, Kontsas K, Siafakas N, Zerva L, Stamboulis E, Kremastinos D, Anastasiou-Nana M. Association of target organ damage with three arterial stiffness indexes according to blood pressure dipping status in untreated hypertensive patients. *Am. J. Hypertens.* 2010; 23:1265-1272.

[18] Hinderliter A L, Blumenthal J A, Waugh R, Chilukuri M, Sherwood A. Ethnic differences in left ventricular structure: relations to hemodynamics and diurnal blood pressure variation. *Am. J. Hypertens.* 2004; 17:43-49.

[19] Middlemost SJ. Beyond systolic and diastolic blood pressure. *Cardiology* 1999: 28-35.

[20] Opie L H. Controversies in cardiology. *Lancet.* 2006; 367:13-14.

[21] Malan L, Malan, NT, Wissing MP, Seedat, YK. 2008. Coping with urbanization: A Cardiometabolic Risk? *Biol. Psych.*, 79:323-328.

[22] Krahn A D, Nguyen-Ho P, Klein G J, Yee R, Skanes A C,Suskin N. QT dispersion: an electrocardiographic derivative of QT prolongation. *Am. Heart J.* 2001; 141:111-116.

[23] Gadaleta F L, Llois S C, Sinisi V A, Quiles J, Avanzas P, Kaski J C. Corrected QT Interval Prolongation: A New Predictor of Cardiovascular Risk in Patients With Non-ST-Elevation Acute Coronary Syndrome. *Rev. Esp. Cardiol. (English Edition)* 2008;61:572-578.

[24] Field J M. *Advanced Cardiovascular Life Support*. Professional Provider Manual Dallas, Am. Heart Ass., 2006.

[25] Kawasaki T, Azuma A, Kuribayashi T, Shiraishi H, Sawada T, Sugihara H, Nakagawa M. Determinant of QT Dispersion in Patients with Hypertrophic Cardiomyopathy. *Pacing & Clin. Electrophys.* 2003; 26:819-826.

[26] Chapman N, Mayet J, Ozkor M, Foale R, Thom S, Poulter N. Ethnic and gender differences in electrocardiographic QT length and QT dispersion in hypertensive subjects. *J. Hum. Hypertens.* 2000;14:403-405.

[27] Krzesinski J M, Rorive G, Van Cauwenberge H. Hypertension and left ventricular hypertrophy. *Acta Cardiol.* 1996;51:143-154.

[28] Loomba R S, Arora R. Hyperglycemia and acute coronary syndrome: a systematic review of hyperglycemia's impact on ACS. *Am. J. Therap.* 2010;17:48-51.

[29] Selvin E, Steffes M W, Zhu H, Matsushita K, Wagenknecht L, Pankow J, Coresh J, Brancati F L. Glycated Hemoglobin, Diabetes, and Cardiovascular Risk in Nondiabetic Adults. *N. Eng. J. Med.* 2010; 362: 800-811.

[30] Demchuk A M, Morgenstern L B, Krieger D W, Linda Chi T, Hu W, Wein T H, Hardy R J, Grotta J C, Buchan A M. Serum glucose level and diabetes predict tissue plasminogen activator-related intracerebral hemorrhage in acute ischemic stroke. *Stroke* 1999;30:34-39.

[31] Gordon T, Kannel W B. Drinking habits and cardiovascular disease: the Framingham Study. *Am. Heart J.* 1983;105:667-673.

[32] Cubrilo-Turek M. Stroke risk factors: recent evidence and new aspects. *Atherosclerosis* 2004; 1262:466-469.

[33] Smith S A. 2010 - Higher "normal" glycated hemoglobin levels were associated with increased risk for diabetes, CVD, stroke, and mortality in adults. *ACP J. Club.* 2010; 53:12-12.

[34] Kulan K, Ural D, Komsuoglu B, Agaçdiken A, Göldeli Ö, Komsuoglu S S. Significance of QTc prolongation on ventricular arrhythmias in patients with left ventricular hypertrophy secondary to essential hypertension. *Int. J. Cardiol.* 1998; 64:179-184.

Reviewed by

Prof YK Seedat, MD (NU Irel); PhD (Med Sci) (hc) (UDW); MD (hc) (Univ Free State); FRSSAf; Hon FCM (SA); FRCP (London); FRCP (Irel); FACP; FCP (SA); FACC, FCCP ; FCPS (PAK); FICA Fellow of the University of KwaZulu Natal, Nelson R Mandela School of Medicine, Faculty of Health Sciences, University of Kwa Zulu Natal. Tel: +27-31 266 1501. Fax: +27-86 665 3216. Email: seedaty1@ukzn.ac.za. Email: seedaty1@gmail.com.

INDEX

#

A

D

E

F

G

H

I

J

K

L

M

N

O

P

T

U

V

W

Y

Z